Software Radio

ANALOG CIRCUITS AND SIGNAL PROCESSING

Series Editors:
Mohammed Ismail. Ohio State University
Momamad Sawan. École Polytechnique de Montréal

For further volumes:
http://www.springer.com/series/7381

Elettra Venosa • fredric j. harris
Francesco A.N. Palmieri

Software Radio

Sampling Rate Selection, Design and Synchronization

 Springer

Elettra Venosa
San Diego State University
San Diego
USA
elven80@virgilio.it

fredric j. harris
San Diego State University
San Diego
USA
fred.harris@sdsu.edu

Francesco A.N. Palmieri
Seconda Università di Napoli
Aversa
Italy
francesco.palmieri@unina2.it

ISBN 978-1-4614-2959-3   ISBN 978-1-4614-0113-1 (eBook)
DOI 10.1007/978-1-4614-0113-1
Springer New York Dordrecht Heidelberg London

Printed on acid-free paper

Springer is part of Springer Science+Business Media (www.springer.com)

"O frati," dissi "che per cento milia
perigli siete giun ti a l'occidente,
a questa tanto picciola vigilia
d'i nostri sensi ch' del rimanente
non vogliate negar l'esperienza,
di retro al sol, del mondo sanza gente.
Considerate la vostra semenza:
fatti non foste a viver come bruti,
ma per seguir virtute e canoscenza."

A. Dante, Inferno, Canto XXVI

Preface

Software Radio (SR) is one of the most important emerging technologies for the future of wireless communication services. By moving radio functionality into software, it promises to give flexible radio systems that are multi-service, multi-standard, multi-band, reconfigurable and reprogrammable by software.

Today's radios are matched to a particular class of signals that are well defined by their carrier frequencies, modulation formats and bandwidths. A radio transmitter today can only up convert signals with well-defined bandwidths over defined center frequencies, while, on the other side of the communication chain, a radio receiver can only down convert well-defined signal bandwidths, transmitted over specified carrier frequencies.

The challenge is to remove such constraints and design systems that can be used as universal platforms that can adapt themselves to specific situations. We would like to have a single transmitter that is able to up convert signals independently of their bandwidth, modulation format and carrier frequency. For every transmission, we would like to detect an available, wide enough, spectral hole which is arbitrarily located in the radio spectrum, and to transmit the signal without changing the hardware platform of the transmitter and receiver devices. In order to achieve this goal, the radio devices have to be software-based and their hardware platform has to be reduced to a minimum. The devices have to be reconfigurable and the signal digitization has to occur as close as possible to the antenna. Such a radio will make possible the dynamic spectrum management ensuring the optimal use of this precious resource.

This book is derived from the Ph.D. thesis of one of the authors, and thus it aggregates the studies, the deductions, the mathematical derivations and the scientific results achieved during a 4-year long period of research activity. It does not have the intention of providing a detailed overview of the current state of the art on software radio technology, nor it is intended as a textbook. It examines specific aspects connected with the realization of software radios discussing some of, what the writers believe to be, the most critical issues connected with it, and thus this book necessarily reflects the biased view of its authors. Specifically the book addresses the following issues: proper low-sampling rate selection in the multi-band received signal scenario, architecture design for both software radio receiver and transmitter devices and radio synchronization. This book also addresses other issues that the authors' recent research work has approached (timing jitter modeling, time interleaved analog to digital converter and spectral analysis).

A scientific approach is followed everywhere which maintains the mathematic formulations to an acceptable level for a reader that has an average signal processing background. This book is aimed at a research community that is interested in further development of efficient algorithm implementations.

The authors tried to do their best for explaining all the concepts as clearly as possible, trying also to avoid losing the details in the description. Many figures and block diagrams have been included for simplifying the understanding process. At the end of every reasoning and mathematical derivation, simulation results are provided in order to confirm the effectiveness of the proposed theories, while many references are given at the end of every chapter for those who would like to enrich their knowledge on the topics.

San Diego, CA   Elettra Venosa
San Diego, CA   fredric j. harris
Aversa, Italy   Francesco A.N. Palmieri

Organization of Book

This book is composed of five main chapters.

Chapter 1 provides the basics on software defined radio (SDR) and discusses its benefits pointing out the current state of the technology. This chapter also clarifies the differences between software defined radio, software radio and adaptive intelligent software radio (AISR), better known as cognitive radio (CR).

Chapter 2 introduces the sampling rate selection issue explaining the key reasons that motivate the use of low-rate sampling techniques in digital signal processor based architectures. This chapter provides a new computational formula for selecting the appropriate low-sampling rate in the case of multi-band signals. Theoretical results are confirmed through Matlab simulations. In this chapter the authors also give a new model for describing the frequency selective effect of the timing jitter affecting the analog to digital converters (ADCs) along with an effective and easy to implement solution for semi-blindly correcting time and gain mismatches in a two-channel time interleaved ADC (TI-ADC) operating on communication signals (Hermitian symmetric signals centered at intermediate frequency).

Chapter 3 faces the problem of designing a software defined radio architecture, both for the transmitter and the receiver. A software defined radio transmitter must be able to simultaneously up convert signals with variable bandwidths and randomly located center frequencies while a software defined radio receiver must be able to simultaneously detect and down convert them. In this chapter the authors present implementable back-end structures for both the SDR transmitter and receiver. The polyphase up and down converter channelizer engines are at the core of the proposed designs. That is the motivation for which this chapter also provides the basics of multirate signal processing with particular attention to the derivation of the standard polyphase down converter channelizer.

In Chap. 4 three octave processors for constant-Q spectrum analyzer are presented. Spectrum analyzers are indispensable for software radios. In fact an adaptive intelligent software radio transmitter must be able to optimize the utilization of the available radio spectrum; in order to achieve this goal it needs to know where the white spaces are located. On the other side, the receiver needs to know the carrier frequencies of the information signals and their bandwidths in order to demodulate them properly.

Chapter 5 faces the synchronization issue in digital radios. After providing an introduction of this topic, this chapter presents a new blind time, frequency and phase synchronization algorithm for a software defined radio receiver operating on quadrature amplitude modulated (QAM) signals. The algorithm is based on information-theoretic criteria.

Contents

List of Figures

List of Tables

Chapter 1
Software Radio: From an Idea to Reality

1.1 Introduction

This chapter offers a general introduction on software radio. It provides the current state of the art on this technology clarifying the differences between software defined radio, software radio and adaptive intelligent software radio (also known as cognitive radio). The chapter also describes the characteristics and the benefits of using software radio technology along with the main issues connected to its design and commercial development. This chapter aims to provide a general picture of all the topics that are covered in this book.

1.2 Software Defined Radio or Software Radio?

The twentieth century saw the explosion of hardware defined radio as a means of communicating all forms of data, audible and visual information over vast distances. These radios have little or no software control. Their structures are fixed in accordance with the applications; the signal modulation formats, the carrier frequencies and bandwidths are only some of the factors that dictate the radio structures. The smallest change to one of these parameters could imply a replacement of the entire radio system.

A consequence of this is, for example, the fact that a television receiver purchased in France does not work in England. The reason, of course, is that the different geographical regions employ different modulation standards for the analog TV as well as for digital TV. Then, the citizens cannot use the same TV for receiving signals in both countries; they need to buy a new television for each country in which they decide to live. Sometimes, even if the communication systems are designed for the same application purposes and they work in the same geographical area, they are not able to communicate between each other. One of the most evident examples

E. Venosa et al., *Software Radio: Sampling Rate Selection, Design and Synchronization*,
Analog Circuits and Signal Processing, DOI 10.1007/978-1-4614-0113-1_1,
© Springer Science+Business Media, LLC 2012

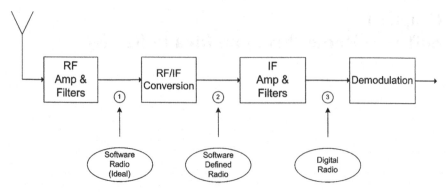

Fig. 1.1 Simple block diagram indicating all the possible places in which the digitization can occur in a radio receiver

of this is that the city police car radio cannot communicate with the city fire truck radio, or with the local hospital ambulance radio even if they have the common purpose of helping and supporting the citizen. Also, the city fire truck radio cannot communicate with the county fire truck radio, or with the radios of the fire truck operated by the adjacent city, or by the state park service, or the international airport. None of these services can communicate with the National Guard, or with the local Port Authority, or with the local Navy base, or the local Coast Guard base, or the US Border Patrol, or US Customs Service.

In an hardware defined radio, if we decide to change one of the parameters of the transmitted signal, like bandwidth or carrier frequency (for example because the carrier frequency we want to use is the only one available at that particular moment), we need to change the transmitter. On the other side, every time we want to receive a signal having different bandwidth or center frequency, we need to change the receiver. Hardware defined transmitter and receiver devices are not flexible at all; we must modify their structure every time we change even one of the transmitting and receiving signal parameters.

In 1991, Joe Mitola coined the term software defined radio. It was referred to a class of reprogrammable (and reconfigurable) devices. At that time it was not clear at which level the digitization should occur to define a radio as software but the concept sounded pretty interesting, and the dream of building a completely reconfigurable radio device involved scientists from all over the world.

Today the exact definition of software defined radio is still controversial, and no consensus exists about the level of reconfigurability needed to qualify a radio as software.

Figure 1.1 shows, in a simple block diagram, all the possible places in which the digitization can occur in a radio receiver. The exact dual example can be portrayed for the radio transmitter. As a reminder we specify that, in the old analog radio, the digitization occurs after the down conversion process (after the signal has been shifted to base-band).

In this book the following classification is used:

- *Software defined radio* is a device in which the digitization of the data is performed, in the receiver, after the wideband filtering, low noise amplification, and down conversion to intermediate frequency (IF) (point 2 in Fig. 1.1). Note that dual processes occur in a software defined radio transmitter.
- *Software radio* is a device in which the digitization occurs, in the receiver, before the intermediate frequency downconversion stage (point 1 in Fig. 1.1). Note that dual processes occur in a software radio transmitter. The term software radio usually denotes a completely reconfigurable radio that can be programmed in software and its physical layer behavior can be significantly altered through software changes.

In short, those two terms differ from each other in the stage in which the signal digitization occurs in both the receiver and the transmitter.

Current radios, often referred to as digital but sometimes referred as software defined radios (depending on their particular structure), after shifting the signals to intermediate frequency, digitize them and assign all the remaining tasks to a digital signal processor. One of the main reasons for shifting the signals to intermediate frequency, before digitizing them, is to reduce their maximum frequency so that a smaller number of samples can be taken for preserving the information content.

Many research issues are still open in designing software defined radio devices. One of the most important issue concerns the selection of proper low-rate sampling frequency in a software radio receiver handling multi-band signals as well as the design of a digital signal processor based receiver structure that is able to simultaneously down convert (up convert, for transmitter side) randomly located center frequency signals having different bandwidths. Also the synchronization issue is still an open topic for such a system. The software radio receiver must be frequency, phase and time synchronized to recover the transmitted signal.

The selection of the proper sampling rate is important because the number of signal samples to be processed directly affects the workload of the radio digital data section. Moreover, sampling speed and precision of the analog to digital converters strongly affect their prices. On the other hand, the issue of designing a structure that is able to simultaneously up convert (and down convert) multiple signals with arbitrary bandwidths over randomly located center frequencies is important in order to avoid digital data section replication in our radios. In this book both these aspects, along with the synchronization issue, are discussed and some efficient solutions are provided. All the theoretical results are supported by Matlab simulations.

Before concluding this section we would like to introduce the reader to the term adaptive intelligent radio, else known as cognitive radio. It is the last stage in the evolution of a software radio. This term refers to a device which is able to adapt itself to the operational environment by automatically adapting, via software, its operational mode.

1.3   Characteristics and Benefits of Software Radio

Implementation of software radios requires digitization at the antenna, allowing complete flexibility in the digital domain. Then it requires both the design of a flexible and efficient digital signal processor based structure and the design of a completely flexible radio frequency front-end for handling a wide range of carrier frequencies, bandwidths and modulation formats. These issues have not been exploited yet in the commercial systems due to technology limitations and cost considerations.

As pointed out in the previous section, in a software defined radio receiver the signals are digitized in intermediate frequency bands. The receiver employs a super heterodyne frequency conversion, in which the radio frequency signals are picked up by the antenna along with other spurious, unwanted signals (noise and interferences), filtered, amplified with a low noise amplifier and mixed with a local oscillator to shift it to intermediate frequency. Depending on the application, the number of stages of this operation may vary. Digitizing the signal in the IF range eliminates the last analog stage in the conventional hardware defined radios in which problems like carrier offset and imaging are encountered. When sampled, digital IF signals give spectral replicas that can be placed accurately near the base-band frequency, allowing frequency translation and digitization to be carried out simultaneously. Digital filtering and sample rate conversion are often needed to interface the output of the ADC to the processing hardware to implement the receiver. Likewise, on the transmitter side, digital filtering and sample rate conversion are often necessary to interface the digital hardware, that creates the modulated waveforms, to the digital to analog converter. Digital signal processing is usually performed using field programmable gate arrays (FPGAs), or application specific integrated circuits (ASICs).

Software defined radio architectures are quite flexible, in the sense that they usually down convert to IF a collection of signals and, after sampling, they are able to shift these signals to base-band via software. Changes in the signal bandwidths and center frequencies are performed by changing some parameters of the digital data section. The flexibility of such a structure can be improved by moving the analog to digital converter closest to the receiver antenna. By digitizing the signals immediately after (and before in the transmitter) the antenna, which is the software radio case, the demodulation (and modulation) processes are performed completely in software and the radio acquires the capability of changing its personality, possibly in real-time, guaranteeing a desired quality of service (QoS). The digitization after the receiver antenna, in fact, allows service providers to upgrade the infrastructure and market new services quickly. It promises multi-functionality, global mobility, ease of manufacture, compactness and power efficiency.

The flexibility in hardware architectures combined with flexibility in software architectures, through the implementation of techniques such as object oriented programming, provides software radio also with the ability to seamlessly integrate itself into multiple networks with widely different air and data interfaces.

1.4 Recap

In this chapter we provided an introduction to software radio technology. We would like to give the readers the idea that the main goal of a modern radio designer is to process the signals as much as he can in the sampled data domain.

The goal today is to move the converters as close as possible to the antennas with the digital signal processing part of the radio handling all the filtering, frequency shifting and signal coding tasks. This guarantees flexibility and the capability of upgrading the radios without changing their hardware structure.

References

1. E. Buracchini, CSELT, "The Software Radio Concept," *IEEE Communications Magazine*, vol. 38, no. 9, 2000.
2. J. Mitola, "The Software Radio Architecture," *IEEE Communications Magazine*, vol. 33, no. 5, 1995.
3. Walter Tuttlebee, *Software Defined Radio Enabling Technologies*, John Wiley and Sons, 2002.
4. Jeffrey H. Reed, *Software Radio, a Modern Approach to Radio Engineering*, Prentice Hall, 2002.
5. B. Wang and K. J. Ray Liu, "Advances in Cognitive Radio Networks: A Survey," *IEEE Journal of Selected Topics in Signal Processing*, vol. 5, no. 1, 2011.
6. A. Sahai, S. M. Mishra, R. Tandra, and K. A. Woyach, "Cognitive Radios for Spectrum Sharing," *IEEE Signal Processing Magazine*, vol. 26, no. 1, 2009.
7. T. Ulversy, "Software Defined Radio: Challenges and Opportunities," *IEEE Communications Survey and Tutorials*, vol. 2, no. 4, 2010.
8. R. Bagheri, A. Mirzaei and M. E. Heidari, "Software-Defined Radio Receiver: Dream to Reality," *IEEE Communication Magazine*, vol. 44, no. 8, 2006.
9. f. harris and W. Lowdermilk, "Software Defined Radio: a Tutorial," *IEEE Instrumentation and Measurement Magazine*, vol. 13, no. 1, 2010.
10. Matthew Sherman, Christian Rodriguez, Ranga Reddy "IEEE Standards Supporting Cognitive Radio and Networks, Dynamic Spectrum Access, and Coexistence," *IEEE Communication Magazine*, vol. 46, no. 7, 2008.
11. Jun Ma, G. Ye Li and Biing Hwang (Fred) Juang, "Signal Processing in Cognitive Radio," *Proceedings of IEEE*, vol. 97, no. 7, 2009.
12. J. Mitola, "Cognitive Radio Architecture Evolution," *Proceedings of IEEE*, vol. 97, no. 4, 2009.
13. R. Tandra, S. M. Mishra, and A. Sahai, "What is a Spectrum Hole and What Does it Take to Recognize One?" *Proceedings of IEEE*, vol. 97, no. 5, 2009.
14. f. j. harris, *Multirate Signal Processing for Communication Systems*, Prentice Hall, Upper Saddle River, New Jersey 07458, 2004.
15. P. P. Vaidyanathan, *Multirate Systems and Filter Banks*, Prentice-Hall, Englewood Cliffs, 1993.

Chapter 2
Sampling Rate Selection, Timing Jitter and Time Interleaved ADCs

2.1 Introduction

A software radio, which receives multiple signals having different bandwidths at randomly located center frequencies, should sample at very low-rates, allowing flexible and efficient handling of multi-standard, multi-band and asynchronous digital data streams.

This chapter addresses the issue of sampling rate selection. After reviewing the standard scheme for down converting and sampling the I-Q components, the authors propose an analysis of a scheme with a single ADC. They derive a general alias-free condition that, if satisfied, guarantees no loss of receiver performance at low-rates in comparison to the analog, or high rate discrete counterparts. The issue is at first addressed for a general expression of the modulated signals and then, a computable formula is provided for accurately checking the alias-free condition in the case of multi-band signals.

In the second section of this chapter, the timing jitter problem, affecting the analog to digital converters, is addressed and a new model for considering its effects in a digital receiver is given. Simulation results are provided for an orthogonal frequency division multiplexing (OFDM) system which is described by using a compact matrix formulation.

In the last section of this chapter the authors provide a blind algorithm for adaptively correcting time and gain mismatches in a two channels time interleaved ADC. Time interleaved analog to digital converters provide a good solution to overcome the limits of the hardware technology. By using more ADCs, working in parallel, the overall sampling rate can be arbitrarily increased. In contrast to the literature on this topic, the proposed solution is applicable for communication scenarios; in fact IF signals have been considered, as an example, in the simulation results presented at the end of the section.

E. Venosa et al., *Software Radio: Sampling Rate Selection, Design and Synchronization*,
Analog Circuits and Signal Processing, DOI 10.1007/978-1-4614-0113-1_2,
© Springer Science+Business Media, LLC 2012

2.2 Sampling Rate Selection

One of the major challenges in the ongoing effort to implement fully digital receivers, is to move the analog to digital converter as close as possible to the receiver antenna. The challenge, of creating the most flexible architectures for the software defined radio, is both theoretical and practical. Samplers have to operate with a large analog bandwidth and all the demodulation tasks (synchronization and decoding) have to be handled by numerical algorithms that can operate in real-time. A software radio should be able to handle superpositions of multi-standard sources operating asynchronously, at different rates and on different bands.

Direct application of Nyquist's sampling theorem to such signals results in very high sampling rates, almost certainly excessive, when compared to their effective information content. The sampling could be performed at much lower rates as it should be possible to reduce it to the actual information rate.

Recently, the signal processing literature has seen a re-emerging interest in sampling theory. Many innovative ideas have been explored by extending more traditional ones (see references at the end of this chapter) and many contributions in the literature refer to non uniform sampling, random or periodic that also via nested decimation, look for the ultimate sampling pattern that can blindly reach Landau's rate.

In this chapter the authors approach the issue from the point of view of uniform sampling, but considering a large class of signals in comparison of the standard low-pass assumption.

Figure 2.1 shows the high level block diagram of a typical digital I-Q down converter with two ADCs. Application of Nyquist sampling theorem to $i(t), q(t)$ components is straightforward after consideration of their maximum frequency. Conversely, Fig. 2.2 depicts the high level block diagram of a software defined radio, this device only uses one ADC and the base-band down conversion is performed digitally.

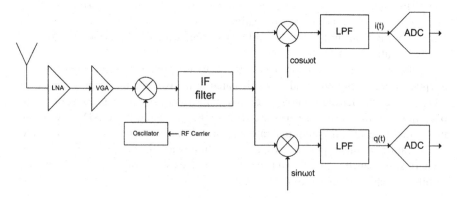

Fig. 2.1 High level block diagram of a conventional I-Q down converter

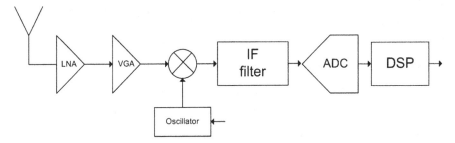

Fig. 2.2 High level block diagram of an ideal software radio receiver

The question here is: can we perform all the operations digitally by using the scheme in Fig. 2.2 and achieve the same performance of Fig. 2.1? If this is true we could move even further, shifting the ADC closer to the antenna.

The objective of this section is to provide the readers with the demonstration that it is theoretically possible to sample, without any information loss, a signal which is generically a multi-band signal avoiding to consider its total bandwidth to be confined below its maximum frequency. If this reasoning is applied, in some cases, it could be possible to decrease the sample rate of the ADCs operating in a fully digital receiver. The following paragraphs are confined to a digital receiver that, equipped with just one ADC, samples at uniform low-rate. The problem that is addressed here, has a different twist with respect to classical sampling theory, because the goal is not the analog signal reconstruction, but rather to provide, after sampling, the sufficient statistics for data decoding.

In the following sections, after reducing the sampling model to a linear problem, the issue of preservation of signal subspace, is addressed. More specifically, we can say that if the sampling rate is not appropriately chosen, symbols may be canceled out with a destructive effect on recovery. The authors analyze this problem, showing that the conditions for conserving the signal space structure after sampling is an alias-free criterion on the mutual spectral correlations of all the functions that make up the analyzed signal. The proposed formula is quite general and it only requires that the functions forming the information signals are incoherent at multiples of the sampling rate in the frequency domain.

Despite the generality of the criterion, which actually suggests further work on the design of more specific signal formats for wideband communications operating with low-rate receivers, the authors apply it to the class of multi-band signals providing a computable formula that allows easy and direct identification of all favorable low-rate sampling frequencies.

2.2.1 Theory and Algorithm for Low-Rate Sampling Frequency Selection

Consider the superposition of N linearly modulated signals

$$y(t) = z(t) + w(t) =$$

$$\sum_{i=1}^{N} z_i(t) + w(t) =$$

$$\sum_{i=1}^{N} \sum_{l=0}^{L_i-1} \sum_{m=1}^{M_i} s_{im}[l]\psi_{im}(t;l) + w(t), \qquad (2.1)$$

where

$$\psi_{im}(t;l) = p_{im}(t - lT_i - t_i)c_{im}(t - t_i),$$

with $s_{im}[l]$ are source symbols, $p_{im}(t)$ are shaping pulses, $c_{im}(t)$ are "carrier" functions, M_i are the number of carriers per signal and L_i are the number of symbols per signal.

The functions $\psi_{im}(t;l)$ are modulated pulses shifted in time at N different symbol rates $\{\frac{1}{T_i}\}$, that account also for the propagation through channels and for the filtering effects. The signals may have different initial times t_i, different carrier frequencies and they can be originated from different sources; $w(t)$ is additive white Gaussian noise (AWGN).

Equation (2.1) represents the most general formula to model the superposition of digitally modulated signals that are asynchronous, may belong to different bands and may have been propagated through different channels. More specifically, for a non dispersive channel, if:

- $N = 1$, $M_1 = 1$, $p_{11}(t) = p(t)$, $c_{11}(t) = \cos 2\pi f_0 t$, where $p(t)$ is an energy pulse, we have pulse amplitude modulation (PAM) with baud-rate $1/T_1$;
- $N = 1$, $M_1 = 2$, $p_{11}(t) = p_{12}(t) = p(t)$, $c_{11}(t) = \cos 2\pi f_0 t$, $c_{12}(t) = \sin 2\pi f_0 t$, we have quadrature amplitude modulation (QAM) with baud-rate $1/T_1$;
- $N = 1$, $p_{1m}(t) = p(t)$, $c_{1m}(t) = \cos 2\pi f_m t$ (m even), $\sin 2\pi f_m t$ (m odd), $m = 1, ..., M$, we have multi-frequency modulation formats such as FMT, OFDM, etc.

The most general case is the superposition of N signals, each expressed as the linear combination of M_i bases (not necessarily orthogonal), built with carriers at various frequencies and with pulses shifted in time at uniformly spaced symbol intervals.

The aim is to study the fully digital receiver that has to decode all $s_{im}[l]$, $i = 1, .., N$; $m = 1, ..., M_i$; $l = 0, ..., L_i - 1$ from (low-rate) uniform samples of $y(t)$ taken at times $t_n = \Delta + nT_{SA}$, where Δ is a fixed ADC delay.

$$y[n] = z[n] + w[n] = z(\Delta + nT_{SA}) + w(\Delta + nT_{SA}) =$$

$$= \sum_{i=1}^{N} z_i(\Delta + nT_{SA}) + w(\Delta + nT_{SA}) =$$

$$= \sum_{i=1}^{N} \sum_{l=0}^{L_i-1} \sum_{m=1}^{M_i} s_{im}[l]\psi_{im}(\Delta + nT_{SA}; l) + w(\Delta + nT_{SA}). \quad (2.2)$$

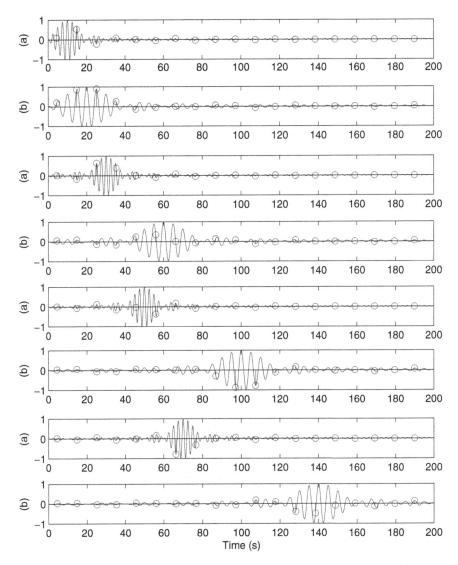

Fig. 2.3 Low-rate sampling bases for $N = 2$. Subplot (**a**): $M_1 = 1$, $p_{11}(t) = sinc(\frac{t}{T_1})$, $c_{11} = cos(2\pi f_{c_1}t)$, $f_{c_1} = \frac{4}{T_1}$ Hz, $T_1 = 10$ s, $t_1 = T_1$, $L_1 = 4$; subplot (**b**): $M_2 = 1$, $p_{21}(t) = sinc(\frac{t}{T_2})$, $c_{21} = cos(2\pi f_{c_2}t)$, $f_{c_2} = \frac{4}{T_2}$ Hz, $T_2 = 20$ s, $t_2 = T_2$, $T_{SA} = 10.8$ s, $M = 10$, and $\Delta = 2.33$ s

The symbol intervals T_i are not necessarily integer multiples of T_{SA}. Therefore sampling and symbol times are totally asynchronous. A real formulation is used in this section to emphasize that the receiver uses only one ADC.

Figure 2.3 shows an example with all the basis functions $\psi_{im}(t; l)$ that make up the superposition of two PAM signals with different symbol rates, different carrier frequencies, and asynchronous sampling. To establish the general conditions for proper sampling, the discrete model must be studied in comparison to the classical optimum receiver that has available the whole analog signal $\{y(t), -\infty < t < \infty\}$.

To simplify notation, the symbol $z(t)$ is used for indicating the superposition of $Q = \sum_{i=1}^{N} M_i L_i$ continuous-time functions

$$y(t) = z(t) + w(t) = s_1\phi_1(t) + \cdots + s_Q\phi_Q(t) + w(t) =$$

$$= [\phi_1(t) \cdots \phi_Q(t)] \begin{bmatrix} s_1 \\ \cdot \\ \cdot \\ \cdot \\ s_Q \end{bmatrix} + w(t) = \phi(t)^T \mathbf{s} + w(t) \qquad (2.3)$$

with

$$\phi_{i+l+m}(t) = \left[\left[[\psi_{im}(t;l)]_{m=1}^{M_i}\right]_{l=1}^{L_i}\right]_{i=1}^{N},$$

$$s_{i+l+m}(t) = \left[\left[[s_{im}[l]]_{m=1}^{M_i}\right]_{l=1}^{L_i}\right]_{i=1}^{N}. \qquad (2.4)$$

Essentially all the continuous time functions that contribute to the multiplex into a unique set have been aligned.

In the discrete-time framework, M samples of $y(t)$ for $n = 0, \ldots, M-1$ can be taken and organized into a vector $\mathbf{y} = (y[0], \ldots, y[M-1])^T$. Via proper choice of Δ and M, a sufficient number of samples are included in the observation window to include essentially all pulse tails for all time shifts. Therefore (2.2) can be reduced to the linear model

$$\mathbf{y} = \mathbf{z} + \mathbf{w} = \mathbf{C}\mathbf{s} + \mathbf{w}, \qquad (2.5)$$

with

$$\mathbf{z} = (z[0], \ldots, z[M-1])^T,$$
$$\mathbf{w} = (w[0], \ldots, w[M-1])^T. \qquad (2.6)$$

Matrix

$$\mathbf{C} = [\phi_1\phi_2...\phi_Q]$$

is $M \times Q$ and contains, in its columns, Q M-dimensional discrete bases for $\mathbf{z}$.

With help of Fig. 2.3, note that even though the analog pulses used for modulation have been obtained by the composition of shifted versions of basic pulses and carriers, the columns of $\mathbf{C}$ *are not* in general shifted versions of a single vector, i.e. the discrete projection space after sampling *is not* shift-invariant, and it *is not* a union of shift-invariant subspaces either.

Note that the sampling process is asynchronous to symbol timing and it may become critical because the discrete-time pattern may vary greatly.

Remember that our aim is not the reconstruction of the analog waveform (this is the most common aim of the people working on sampling theory). Rather, we want to directly find the conditions for which a fully digital receiver has the same information as an optimal analog receiver to decode **s**. Therefore the contents of this section focus on:

- How the discrete basis matrix $\mathbf{C} = [\phi_1\phi_2...\phi_Q]$ is structured as a consequence of sampling in terms of dimensions, rank and relations with its continuous-time counterpart;
- How the noise after sampling affects the decoding process.

The reasoning starts with the following

Definition 2.1. A set of basis sequences $\{\phi_1[n], \phi_2[n]...\phi_Q[n]\}$, obtained by sampling $\{\phi_1(t), ..., \phi_Q(t)\}$ at rate $f_{SA} = 1/T_{SA}$, is called *alias-free* if

$$\sum_{n=-\infty}^{+\infty} \int \Phi_p(f)\Phi_q^*\left(f - \frac{m}{T_{SA}}\right) df = \mathcal{F}[\phi_p(t)\phi_q(t)]_{f=mf_{SA}} = 0,$$

$$m \neq 0, \quad \forall\, p, q = 1, ..., Q, \tag{2.7}$$

where $\Phi_p(f)$ and $\Phi_q(f)$ are the Fourier transforms of $\phi_p(t)$ and $\phi_q(t)$ respectively.

From the alias-free condition (2.7) the following theorem can be stated

Theorem 2.1. *For an alias-free set* $\{\phi_1[n], \phi_2[n], ..., \phi_Q[n]\}$, *obtained by sampling* $\{\phi_1(t), ..., \phi_Q(t)\}$ *at rate* $f_{SA} = 1/T_{SA}$,

$$\sum_{n=-\infty}^{\infty} \phi_p[n]\phi_q[n] = \frac{1}{T_{SA}} \int_{-\infty}^{\infty} \phi_p(t)\phi_q(t)dt,$$

$$\forall\, p, q = 1, ..., Q. \tag{2.8}$$

Note that to have alias-free bases means to have, in the discrete-time domain, the same vector space that is available in the continuous-time domain. All the cross products and the energies (scaled) are preserved and available for optimal receiver design.

Proof. Sampling the energy bases $\phi_1(t), ..., \phi_Q(t)$ at times $t_n = nT_{SA} + \Delta$, $n = -\infty, ..., \infty$ and arbitrary Δ,

$$\sum_{n=-\infty}^{\infty} \phi_p[n]\phi_q[n] =$$

$$\sum_{n=-\infty}^{\infty} \phi_p(nT_{SA} + \Delta)\phi_q(nT_{SA} + \Delta) =$$

$$\sum_{n=-\infty}^{\infty} \int_{-\infty}^{\infty} \int_{-\infty}^{\infty} \Phi_p(\xi)\Phi_q^*(\eta)e^{j2\pi(\xi-\eta)(nT_{SA}+\Delta)}d\xi d\eta =$$

$$\int_{-\infty}^{\infty} \int_{-\infty}^{\infty} \Phi_p(\xi)\Phi_q^*(\eta)e^{j2\pi(\xi-\eta)\Delta}\frac{1}{T_{SA}}\sum_{n=-\infty}^{\infty}\delta\left(\xi-\eta-\frac{n}{T_{SA}}\right)d\xi d\eta =$$

$$\frac{1}{T_{SA}}\sum_{n=-\infty}^{\infty}e^{j2\pi\Delta\frac{n}{T_{SA}}}\int_{-\infty}^{\infty}\Phi_p(f)\Phi_q^*\left(f-\frac{n}{T_{SA}}\right)df,$$

$$\forall\, p,q = 1,...,Q. \tag{2.9}$$

Expression (2.9) is periodic in Δ and if

$$\int_{-\infty}^{\infty}\Phi_p(f)\Phi_q^*\left(f-\frac{n}{T_{SA}}\right)df = 0, \quad n \neq 0, \tag{2.10}$$

the equivalence

$$\sum_{n=-\infty}^{\infty}\phi_p[n]\phi_q[n] = \frac{1}{T_{SA}}\int_{-\infty}^{\infty}\Phi_p(f)\Phi_q^*(f)df, \tag{2.11}$$

that gives (2.8) is valid. ▷

Note that, in the theorem, the authors have considered infinite-length basis sequences. However, by choosing M sufficiently large, and Δ, to include essentially all the tails of our energy signals, the following, arbitrary approximation, can be written

$$\sum_{n=-\infty}^{\infty}\phi_p(nT_{SA}+\Delta)\phi_q(nT_{SA}+\Delta) \simeq$$

$$\simeq \sum_{n=0}^{M-1}\phi_p(nT_{SA}+\Delta)\phi_q(nT_{SA}+\Delta) = \phi_p^T\phi_q. \tag{2.12}$$

Therefore, in alias-free condition it is

$$C^T C = \frac{1}{T_{SA}}R_a, \tag{2.13}$$

where $R_a = \left[\int \phi_p(t)\phi_q(t)dt\right]_{pq}$ is the "analog" bases' correlation matrix.

It can be easily verified that:

For any pair of modulating symbol vectors s and u the Euclidean distance between z(t)|s and z(t)|u simply scales with T_{SA} to the Euclidean distance between z[n]|s and z[n]|u, if the alias-free condition is satisfied.

Proof. The squared distance between the modulated analog signals is

$$d_a^2 = \int ([z(t)|\mathbf{s}] - [z(t)|\mathbf{u}])^2 dt = \mathbf{s}^T R_a \mathbf{s} + \mathbf{u}^T R_a \mathbf{u} - 2\mathbf{s}^T R_a \mathbf{u}; \qquad (2.14)$$

the squared distance between the sampled versions is

$$d^2 = \sum_n ([z[n]|\mathbf{s}] - [z[n]|\mathbf{u}])^2 =$$
$$= \mathbf{s}^T C^T C \mathbf{s} + \mathbf{u}^T C^T C \mathbf{u} - 2\mathbf{s}^T C^T C \mathbf{u}. \qquad (2.15)$$

If the alias-free condition is satisfied:

$$C^T C = \frac{1}{T_{SA}} R_a$$

and

$$d^2 = \frac{1}{T_{SA}} d_a^2$$

i.e. after sampling, in the new vector space the symbol topology does not change. ▷

The alias-free condition (2.7) is then a compact general criterion for designing a discrete receiver with the guarantee that sampling does not affect receiver performance in comparison to a receiver built from the analog basis functions.

The traditional approach to digital receiver design has been to assume that the basis functions are well sampled in accordance with a reconstruction criterion. Typically they are assumed low-pass and sampled at sufficiently high rates. In this more general framework, we can check the alias-free condition (2.7) and sample signals that are neither considered low-pass nor necessarily band-limited.

Without worrying about analog signal reconstruction we can directly focus on the preservation of the scalar products. In this way all the properties of the analog bases are translated to the discrete framework. For example, if the analog basis functions are linearly independent and the alias-free condition is satisfied, then the discrete bases are linearly independent and the discrete time correlation matrix $R_{T_{SA}} = C^T C$ is non singular.

The alias-free condition is essentially an *incoherence* condition in the frequency domain. It requires mutual spectral correlations to be nullified at multiples of the sampling frequency. Incoherence conditions on dictionaries have been used also in the context of compressive sensing. However sampling rate choice for receiver design following the alias-free condition (2.7) has never been presented in the literature.

The alias-free criterion can also be equivalently applied to any linear space obtained from the original set $\phi(t) = \{\phi_1(t), ..., \phi_Q(t)\}$. For example, if the rank of $\phi(t)$ is less than Q, we can use a new representation basis (orthogonal or non-orthogonal) $g(t) = \{g_1(t), ..., g_r(t)\}$ on which to project the signal, and apply all the considerations on the new basis set.

In alias-free condition, optimal receivers can be designed at low-rates standard techniques with no loss of performance with respect to their continuous-time, or high rate counterparts.

Before considering the receiver design problem and the role of noise aliasing in the low-rate sampling scenario, allow us to conclude that all the issue is now reduced to finding for a given communication scenario, sampling rates that satisfy the alias-free conditions (2.7). Also, the search for functions that satisfy the alias-free conditions (2.7), essentially constrained on the nulls of their spectral auto- and mutual-correlations, opens new opportunities in the research of more appropriate signal formats for low-rate sampling.

2.2.2 Multi-Band Signals Case

In this section, the alias-free criterion is applied to the typical class of band-pass bases that belong to multiple separate bands. It is shown that the scalar product preservation boils down to a more traditional alias-free request for non overlapping bands after periodic replication.

Even though in the following an original formula for explicitly computing all the allowed sampling rates is proposed, the connection to the signal space preservation is relevant for possibly claiming optimality of decoders that are based on the low-rate discrete-time sequences.

In the multi-band scenario, imperfect knowledge of the basis function can allow accurate sampling rate choice, because only knowledge of the signal bands is necessary. Also, sampling rate inaccuracies and imperfect knowledge of the signal bands, can be easily accounted for with guard bands.

Assume for simplicity that the problem has been reduced to L basis functions, each one belonging to a different frequency set

$$\mathcal{B}_p = [-b_p, -a_p] \cup [a_p, b_p], \quad p = 1, \dots, L. \tag{2.16}$$

Therefore the signal belongs to the frequency set

$$\mathcal{B}_z = [-b_L, -a_L] \cup \cdots \cup [-b_1, -a_1] \cup [a_1, b_1] \cdots \cup [a_L, b_L]. \tag{2.17}$$

If no constraints are imposed on the signal spectra within each band, it is possible to associate to each function $\Phi_p(f)$ an indicator $I_p(f)$ that is one the frequency support and zero elsewhere. Hence a sufficient condition for alias-free sampling is

$$\int_{-\infty}^{+\infty} I_p(f) I_q(f - \frac{m}{T_{SA}}) df = \mathcal{F}[i_p(t) i_q(t)]_{f=m f_{SA}} = 0, \tag{2.18}$$

$$m \neq 0, \quad \forall\, p, q = 1, \dots, Q,$$

where $I_p(f) = \mathcal{F}[i_p(t)]$.

The above expression measures the amount of overlap between $I_p(f)$ and $I_q(f - \frac{m}{T_{SA}})$ and it is ≥ 0.

Therefore, instead of imposing separate $L(L + 1)/2$ conditions, we can more compactly require that

$$\mathcal{F}\left[\left(\sum_{p=1}^{L} i_p(t)\right)^2\right]_{f=mf_{SA}} = 0, \quad \forall \, m \neq 0. \tag{2.19}$$

By defining

$$I_z(f) = \mathcal{F}[i_z(t)] = \sum_{p=1}^{L} I_p(f), \tag{2.20}$$

the alias-free condition becomes

$$\int_{-\infty}^{+\infty} I_z(f) I_z\left(f - \frac{m}{T_{SA}}\right) df = \mathcal{F}[i_z^2(t)]_{f=mf_{SA}} = 0, \quad m \neq 0. \tag{2.21}$$

Formula (2.21) is a sufficient condition from the general alias-free condition (2.7). It assumes a more familiar form as a check for the absence of spectral overlaps. Essentially (2.21) means that no integer multiples of f_{SA} can fall in $\mathcal{B}_{i_z^2}$, the frequency support of $i_z^2(t)$. It is clear that the connection with (2.7) ensures that the absence of spectral overlaps is a certain opportunity for optimal receiver design, also in the low-rate regime.

In the multi-band scenario, no closed form equation exists, until now, for deciding the appropriate sampling rates as the allowed ones are distributed with very irregular patterns. The band-pass sampling theorem well describes the single band case. However an SDR designer could have the benefit of greater flexibility if he could choose from the *whole* set of allowed sampling rates in the multi-band scenario. In the following an explicit computable formula for deriving the allowed low-rate sampling frequencies in the general multi-band signal case is proposed.

We write the indicator function $I_z(f)$ as

$$I_z(f) = \sum_{i=1}^{4L} (-1)^{i+1} u(f - f_i), \tag{2.22}$$

where $u(x)$ is the unit step function and the $4L$ *node frequencies* f_i are

$$f_{2k-1} = \begin{cases} -b_{L-k+1} & k = 1, \dots, L \\ a_{k-L} & k = L+1, \dots, 2L \end{cases}$$

$$f_{2k} = \begin{cases} -a_{L-k+1} & k = 1, \dots, L \\ b_{k-L} & k = L+1, \dots, 2L \end{cases} \tag{2.23}$$

by convolving, we obtain

$$(I_z * I_z)(f) =$$

$$\sum_{i=1}^{4L}\sum_{j=1}^{4L}(-1)^{i+1}(-1)^{j+1}\int_{-\infty}^{\infty}u(\nu - f_i)u(f - \nu - f_j)d\nu =$$

$$\sum_{i=1}^{4L}\sum_{j=1}^{4L}(-1)^{i+j}r\left(f - (f_i + f_j)\right), \tag{2.24}$$

where $r(x) = xu(x)$ is the ramp function.

Therefore, *the multi-band signal is properly sampled at frequency f_{SA} if*

$$u\left(\sum_{i=1}^{4L}\sum_{j=1}^{4L}(-1)^{i+j}r\left(nf_{SA} - (f_i + f_j)\right)\right) = 0,$$

$$n = 1, 2, \ldots, \left\lceil \frac{2b_L}{f_{SA}} \right\rceil. \tag{2.25}$$

Condition (2.21) and its computational formula (2.25) are completely general and include all the well known results for low-pass and band-pass signals. More importantly, they allow precise low-rate selection for any multi-band signal. More specifically:

- For a *low-pass signal*, $\mathcal{B}_z = [-b, b]$ and $\mathcal{B}_{iz^2} = [-2b, 2b]$. Therefore $f_{SA} > 2b$, i.e. the classical Nyquist's sampling rate satisfies the criterion;
- For a *band-pass signal*, $\mathcal{B}_z = [-b, -a] \cup [a, b]$ with $b - a = W$; then $\mathcal{B}_{iz^2} = [-2b, -2a] \cup [-W, W] \cup [2a, 2b]$. i.e. no aliasing occurs if $2b \leq f_{SA} \leq \infty$, $\frac{2b}{2} \leq f_{SA} \leq 2a$, $\frac{2b}{3} \leq f_{SA} \leq \frac{2a}{2}$ until $\frac{2b}{k} \leq f_{SA} \leq \frac{2a}{k-1}$ with $1 \leq k \leq \lfloor \frac{b}{W} \rfloor$. These are the conditions of the classical band-pass sampling theorem;
- For a *multi-band signal*, $\mathcal{B}_z$ is (2.17) and $\mathcal{B}_{iz^2}$ has a much more complicated structure and the proper sampling frequency must be selected with the help of computational formula (2.25).

As an example, a multi-band signal in the bands $\mathcal{B}_1 = [a_1, b_1] = [25, 30]\,\text{kHz}$, $\mathcal{B}_2 = [a_2, b_2] = [36, 38]\,\text{kHz}$, $\mathcal{B}_3 = [a_3, b_3] = [43, 45]\,\text{kHz}$ is considered. Figure 2.4 shows a plot of all the allowed sampling rates computed by applying formula (2.25). The plot also shows, with the help of two arrows, the allowed sampling rates if the signal is considered globally in the band $[a_1, b_3] = [25, 45]\,\text{kHz}$ (band-pass signal). By using the computational formula (2.25) we find many more (all) allowed rates, because it accounts also for all the spectral holes left open during spectral replications.

To supply the designer with further details, in the following an expression that computes all the frequency ranges after sampling in the normalized frequency

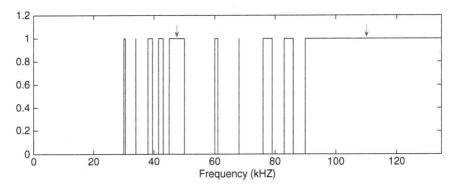

Fig. 2.4 Allowed sampling rates for a three-band signal in the bands $\mathcal{B}_1 = [a_1, b_1] = [25, 30]$ kHz, $\mathcal{B}_2 = [a_2, b_2] = [36, 38]$ kHz, $\mathcal{B}_3 = [a_3, b_3] = [43, 45]$ kHz. The square pulses indicate, on the x axis, the allowed sampling rates calculated by using (2.25). The arrows indicate the allowed sampling rate intervals if the signal is considered globally in the band $[a_1, b_3] = [25, 45]$ kHz (band-pass sampling theorem)

domain $\nu = f/f_{SA}$ is provided. Recall that the digital spectrum is hermitian and periodic with period equal to one. Therefore you can confine your attention to the interval $\nu \in [0, \frac{1}{2}]$.

For a generic band-pass component in $[a_i, b_i]$ Hz, define a *decimation factor* k_i as

$$k_i = \left\lceil \frac{2b_i}{f_{SA}} \right\rceil. \tag{2.26}$$

If k_i is odd, the spectral component after alias-free sampling (of the whole multi-band signal) is located (not inverted) in the interval

$$0 \leq \left[\frac{a_i}{f_{SA}} - \frac{k_i - 1}{2}, \frac{b_i}{f_{SA}} - \frac{k_i - 1}{2} \right] \leq \frac{1}{2}. \tag{2.27}$$

If k_i is even, the spectral component after alias-free sampling (of the whole multi-band signal) is located (inverted) in the interval

$$0 \leq \left[-\frac{b_i}{f_{SA}} + \frac{k_i}{2}, -\frac{a_i}{f_{SA}} + \frac{k_i}{2} \right] \leq \frac{1}{2}. \tag{2.28}$$

Figure 2.5 shows the sampling effects in the discrete frequency domain of the three-band signal of Fig. 2.4. More specifically:

- For $f_{SA} = 91$ kHz (Nyquist Rate) the signal replicas are in the interval $[.2747, .3297] \cup [.3956, .4176] \cup [.4725, .4945]$;
- For $f_{SA} = 47$ kHz the signal replicas are in the interval $[.3617, .4681] \cup [.1915, .2340] \cup [.0426, .0851]$;
- For $f_{SA} = 42$ kHz the signal replicas are in the interval $[.2857, .4048] \cup [.0952, .1429] \cup [.0238, .0714]$;

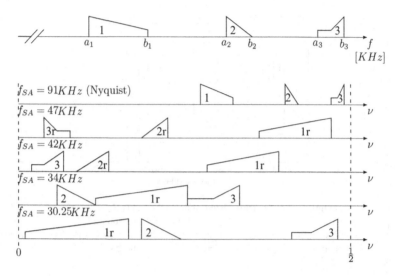

Fig. 2.5 Discrete spectra after alias-free sampling for the example of Fig. 2.4. The first plot shows the signal spectrum before sampling (in kHz). The other plots show the resulting spectra after sampling in the normalized frequency domain with ranges computed with (2.27) and (2.28)

- For $f_{SA} = 34\,\text{kHz}$ the signal replicas are in the interval $[.1176, .2647] \cup [.0588, .1176] \cup [.2647, .3235]$;
- For $f_{SA} = 30.25\,\text{kHz}$ the signal replicas are in the interval $[.0083, .1736] \cup [.1901, .2562] \cup [.4215, .4876]$.

Note that some spectral components may be inverted and/or interlaced in different orders according to k_i, which may be different for every band and for every sampling rate.

2.2.3 Noise Aliasing

The presence of noise in (2.1) and (2.3) requires consideration of what happens, after sampling, in (2.2) and (2.5). We want to underline the effects of noise on the receiver design in the low-rate sampling scenario. Intuitively, by choosing a low sampling frequency, we take fewer samples of the information signal but we also account for less noise. Moreover the reasoning is not so trivial and such scaling needs to be analyzed carefully with reference to the two main sources of noise in the SDR receiver:

- The channel, with noise $w_c(t)$ that is received by the system with the information signal; it may be confined to specific bands and be colored;
- The "open band" of the ADC, with a wideband noise $w_{ADC}(t)$ that is added to the received signal.

In the following section both the noise contributions to $w(t) = w_c(t) + w_{ADC}(t)$ are specifically analyzed.

Channel Noise: If the analog channel noise has autocorrelation $R_{aw_c}(\tau) = E[w_c(t)w_c(t - \tau)]$, after sampling $w_c[n]$ has autocorrelation

$$R_{w_c}[l] = E[w_c[n]w_c[n - l]] = R_{aw_c}(nT_{SA}). \tag{2.29}$$

Note that a large finite-length observation window has been considered. Assuming T to be the total length and by using a sampling interval T_{SA} we have $T = MT_{SA}$.

The total noise energy for the analog signal is

$$\mathcal{E}_a = \int_T E[w^2(t)]dt = \sigma_w^2 T \tag{2.30}$$

where $\sigma_w^2 = R_{aw}(0)$.

After sampling the total energy for the discrete signal is

$$\mathcal{E} = \sum_{n=0}^{M-1} E[w^2[n]] = M\sigma_w^2 = \frac{\mathcal{E}_a}{T_{SA}}. \tag{2.31}$$

Hence, after sampling noise scales with the sampling frequency, just like the signal energy in alias-free conditions. Therefore, any receiver designed on the discrete-time sequence observes the same vector space as its analog counterpart if the alias-free conditions are satisfied.

In alias-free condition any receiver that can be designed on $y(t)$ can be equivalently designed on $y[n]$.

ADC Noise: In addition to proper sampling rate selection, one technical difficulty that has limited the widespread use of low-rate direct sampling is the noise coming from the hardware devices and in particular from the ADC. Even if the converter samples at low-rate, it must "see" a large input bandwidth that causes device noise to enter the system.

Device producers are currently working to supply the market with ADCs that have large bandwidths with good noise figures, but we have to realize that after low rate sampling, and consequent spectral folding, ADC noise can become a penalizing factor in receiver performance.

The standard simplified model for the noisy ADC is shown in Fig. 2.6. The low-pass filter at the input of the ADC must have a very large analog band, $\mathcal{B}_{ADC} = [-f_{ADC}, f_{ADC}]$, at least sufficient to contain frequencies up to the maximum frequency of the incoming signal, i.e. $f_{ADC} > b_Q$. In fact, even if the information signal and the channel noise are limited to specific bands, the sampler that works at low-rate on a high-frequency signal must have a sufficient time resolution to capture the fine signal details. Such an "open band" at the input causes extra noise to enter the receiver.

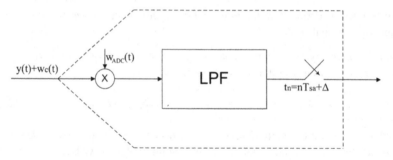

Fig. 2.6 Model of the analog to digital converter with the device noise

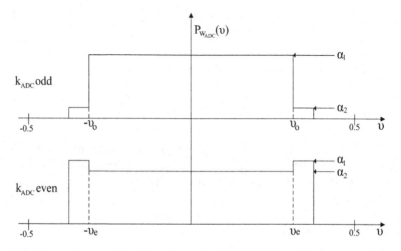

Fig. 2.7 Noise power spectral density at the output of the ADC with analog bandwidth f_{ADC} and sampling rate f_{SA} for odd and even decimation factor k_{ADC}; $\alpha_1 = (K_{ADC} - 1)f_{SA}\frac{\eta_{ADC}}{2}$, $\alpha_2 = K_{ADC}f_{SA}\frac{\eta_{ADC}}{2}$

The noise power spectral density (in power/Hz), initially negligible, increases with undersampling because of repeated spectral folding (noise aliasing).

Modeling the device noise as flat low-pass Gaussian noise with power spectral density $\frac{\eta_{ADC}}{2}$ and defining the ADC decimation factor as

$$k_{ADC} = \left\lceil \frac{2f_{ADC}}{f_{SA}} \right\rceil \geq 1, \tag{2.32}$$

by repeated folding of the low-pass spectrum of $w_{ADC}(t)$, it is easy to obtain the noise power spectral densities (PSDs) shown in Fig. 2.7 with

$$\nu_e = \frac{k_{ADC}}{2} - \frac{f_{ADC}}{f_{SA}}, \quad \nu_o = -\frac{k_{ADC} - 1}{2} + \frac{f_{ADC}}{f_{SA}}. \tag{2.33}$$

The autocorrelation

$$R_{w_{ADC}}[m] = \int_{-\frac{1}{2}}^{\frac{1}{2}} P_{w_{ADC}}(\nu)e^{j2\pi m\nu}d\nu \tag{2.34}$$

from Fig. 2.7 is

$$R_{w_{ADC}}[m] = \begin{cases} \frac{\eta_{ADC}}{T_{SA}}\left(\frac{k_{ADC}-1}{2}\delta[m] + \nu_o \text{sinc}2\nu_o m\right), & k_{ADC} \text{ odd,} \\ \frac{\eta_{ADC}}{T_{SA}}\left(\frac{k_{ADC}}{2}\delta[m] - \nu_e \text{sinc}2\nu_e m\right), & k_{ADC} \text{ even.} \end{cases} \tag{2.35}$$

Note, with the help of Fig. 2.7, that the noise aliasing is weakly correlated when k_{ADC} is large, in fact, in this case the steps on both power spectral densities become negligible. Therefore, it is reasonable to assume that the ADC noise is white with variance

$$\sigma_{ADC}^2 = \frac{\eta_{ADC}}{T_{SA}}\left(\frac{2k_{ADC}-1}{4}\right). \tag{2.36}$$

Unfortunately, as pointed out above, the noise contribution from the ADC grows with k_{ADC} becoming significant after low-rate sampling.

The receiver designer, from knowledge of η_{ADC}, which is typically very small, and the decimation factor k_{ADC}, can account for worsening of the performance. In fact the new variance ratio is

$$\frac{\sigma_s^2}{\sigma_c^2 + \sigma_{ADC}^2} = \frac{\sigma_s^2}{\sigma_c^2}\frac{1}{1 + \frac{\sigma_{ADC}^2}{\sigma_c^2}} \tag{2.37}$$

and the designer can shift his operating point in the performance curves by reducing the signal to noise ratio by

$$\Delta_{ADC} = 10\log_{10}\left(1 + \frac{\sigma_{ADC}^2}{\sigma_c^2}\right) \quad dB \tag{2.38}$$

2.2.4 Simulation Results

Practical receivers are usually based on various combinations of filters and decoders, corresponding to the modulation formats that compose the multi-band signal $z(t)$. A theoretical receiver model is considered in this section with the aim of demonstrating that, if properly chosen, the selection of a low-rate sampling frequency does not affect the performance.

The basis functions $\psi_{im}(t;l)$ must satisfy the alias-free criterion and, for this reason, they should be known. However, for multi-band signals, it is enough that just the signal bands are known, at least approximately, so that sampling can be performed in a conservative way causing no loss of information. Blind or semi-blind techniques can be used to estimate the signal bases in $\mathbf{C}$ for receiver design.

The multiple possibilities that a designer has in choosing receiver structures, both for accuracy and computational complexity, are beyond the scope of this book. The authors refer here only to the solution of the general compact linear model (2.5) showing, using few examples, how an appropriate sampling rate choice is crucial for obtaining optimality in receiver performance.

Note that the dimension of the problem varies with the sampling rate even though we assume that M is sufficiently large to include essentially all pulse tails in our observation window. When the information signals are tightly packed in time it could happen that the dimension M is exceeded by the degrees of freedom Q; more specifically various cases can be distinguished:

- When $M > Q$, the number of observations exceeds the degrees of freedom Q the linear problem is so-called *underloaded*, or *overdetermined*;
- When $M = Q$, the problem is *fully loaded*, or *fully determined*;
- When $M < Q$, the problem is *overloaded*, or *underdetermined* (borrowing the terminology from the CDMA literature).

In all the cases, it is well known that under the assumption of white Gaussian noise $f_{\mathbf{w}}(\mathbf{w}) = \mathcal{N}(\mathbf{w}; \mathbf{0}, \sigma_w^2 I_M)$, the optimal solution is the maximum likelihood (ML) detector,

$$\hat{\mathbf{s}}_{ML} = argmax_{\mathbf{s}} e^{-\frac{1}{2\sigma_w^2}||\mathbf{y}-\mathbf{Cs}||^2} = argmin_{\mathbf{s}}||\mathbf{y} - \mathbf{Cs}||^2. \qquad (2.39)$$

Despite the simple formulation, the ML solution has exponential complexity and can be applied directly only to low-dimensional problems.

Alternatively, many sub-optimal receivers have been proposed in the literature, the most common one being the minimum mean square error (MMSE) receiver based on the linear estimate

$$\hat{\mathbf{s}}_{MMSE} = R_s \mathbf{C}^T \left(\mathbf{C} R_s \mathbf{C}^T + \sigma_w^2 I_N \right)^{-1} \mathbf{y}, \qquad (2.40)$$

with $R_s = E[\mathbf{s}\mathbf{s}^T] = diag(\mathcal{E}_1, ..., \mathcal{E}_Q)$, and on separate decisions on each component. Equation (2.40) is derived under the assumption of zero-mean and separable constellations. Note that, in overloaded problems, the MMSE solution usually does not perform well and more complex iterative algorithms must be used.

In the following simulations, the superposition of three pulse amplitude modulated signals ($N = 3$, $M_1 = M_2 = M_3 = 1$) that belong to three different bands have been considered; they are exactly bandlimited, have different symbol rates and are asynchronous. For each PAM signal 5 symbols ($L_1 = L_2 = L_3 = 5$) have been taken; more specifically:

$$\psi_{11}(t;l) = \text{sinc}\frac{1}{T_1}(t - lT_1 - t_1)\cos 2\pi f_1(t - t_1),$$

$$\psi_{21}(t;l) = \text{sinc}\frac{1}{T_2}(t - lT_2 - t_2)\cos 2\pi f_2(t - t_2),$$

$$\psi_{31}(t;l) = \text{sinc}\frac{1}{T_3}(t - lT_3 - t_3)\cos 2\pi f_3(t - t_3),$$

with $l = 0, ..., 4$; $f_1 = 15\,\text{kHz}$; $t_1 = 1.5\,\text{ms}$; $T_1 = 2\,\text{ms}$; $f_2 = 17\,\text{kHz}$; $t_2 = 1.1\,\text{ms}$; $T_2 = 2\,\text{ms}$; $f_3 = 18\,\text{kHz}$; $t_3 = 0.3\,\text{ms}$; $T_3 = 1\,\text{ms}$.
The corresponding bands are:

$$[a_1, b_1] = [14.75, 15.25]\,\text{kHz},$$

$$[a_2, b_2] = [16.75, 17.25]\,\text{kHz},$$

$$[a_3, b_3] = [17.5, 18.5]\,\text{kHz}.$$

The observation window starts at $t = -3.5\,\text{ms}$ and ends at $t = 16\,\text{ms}$ including essentially all the tails on both sides. Therefore, according to the sampling model, $\Delta = -3.5\,\text{ms}$ and $M = \lfloor 19.5/T_{SA} \rfloor$.

Note that the maximum frequency is 18.5 kHz. Therefore the Nyquist criterion would require $T_{SA} < \frac{1}{37} = 0.0270\,\text{ms}$. However lower rates satisfy the alias-free condition.

The allowed sampling intervals are shown in the lower subplot of Fig. 2.8 by the highest indicator function. The shorter indicator function represents the allowed sampling intervals if the signal is considered globally in the band $[a_1, b_3] = [14.75, 18.5]\,\text{kHz}$ (essentially ignoring the gaps between the bands). The same Fig. 2.8 shows the condition number of $\mathbf{C}^T\mathbf{C}$ for varying sampling intervals.

Note that the analog bases are linearly independent and, under the alias-free condition, the sampled bases should remain non singular. In fact the allowed rates coincide with the low condition numbers confirming that improper sampling destroys the signal space structure. Note also that the number of samples M changes according to the sampling rate, i.e. the problem order changes with the sampling rate.

In this simulation the total number of samples M always exceeds the total number of unknown components $3 \times 5 = 15$ (for example: for $T_{SA} = 0.02\,\text{ms}$, $M = 975$; for $T_{SA} = 0.16\,\text{ms}$, $M = 121$, for $T_{SA} = 1\,\text{ms}$, $M = 19$).

Figure 2.9 shows the results of the simulation of the data stream decoding. The whole complex of 15 symbols has been presented to the system 500 times and the bit error rate (BER) has been estimated. The author reports the results for an ML receiver as well as for an MMSE receiver at three different sampling rates: $f_{SA} = 40\,\text{kHz} = 1/(0.0250\,\text{ms})$ (Nyquist), $f_{SA} = 16.038\,\text{kHz} = 1/(0.0624\,\text{ms})$ (not allowed) and $f_{SA} = 7.75\,\text{kHz} = 1/(0.129\,\text{ms})$ (allowed). The simulation clearly

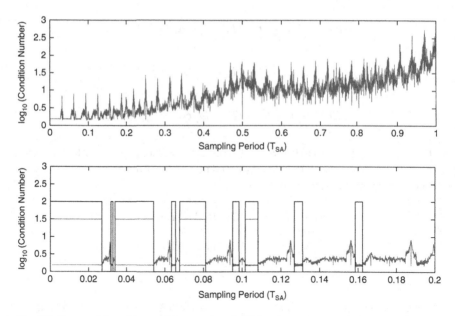

Fig. 2.8 Superposition of the condition number of $\mathbf{C}^T\mathbf{C}$ on log scale at varying sampling intervals and the indicator functions when the alias-free sampling is applied on the superposition of three PAM signals. The higher indicator functions correspond to the general alias-free sampling application while the lower indicator functions represent the allowed sampling intervals when the three signal bands are considered globally as only one

confirms that the performance at the Nyquist rate are indistinguishable with those at a much lower sampling rate. It also clearly shows loss of performance when the alias-free criterion is not satisfied. Note that the "not allowed" rate is higher than the allowed one.

Many more simulations have been performed using signals composed of more bands, also implementing the corresponding optimal receiver and verifying the probability of error. The author also ran simulations with non-exactly bandlimited signals with rectangular, triangular and other pulse shapes. The results are consistently robust with respect to these cases, even with an approximate bandwidth estimation.

2.3 Timing Jitter

In designing a fully digital receiver, the precision of the sampling process is a central issue. Clock instabilities, viewed as timing jitter, may cause performance loss, especially in wideband modulation formats. In this chapter, the authors propose a new general model for describing the destructive effects of the timing jitter. The random jitter can be modeled as a time-varying impulse response with

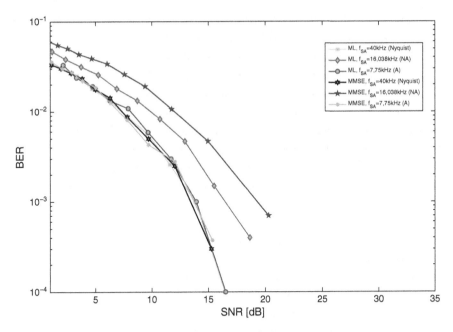

Fig. 2.9 Bit error rate for three superimposed PAM signals at three different sampling rates. When the alias-free condition is satisfied the receivers designed on the sparsely sampled signal perform just like the same receivers designed at Nyquist's rate. Note that the improper sampling rate is higher than the allowed one (A = allowed: NA = not allowed)

complementary coefficients (that sum up to one) and here it is demonstrated how a simplified model that uses only three coefficients of the impulse response captures the essence of the process with frequency selective effects.

The three samples model is then applied to a standard OFDM transceiver chain, where all the calculations are presented in a compact matrix form. The theoretical results are, at the end of the section, confirmed by simulations.

2.3.1 Jittered Sampling

We propose here a simplified formulation in which the role of the misalignment of the local oscillator is included.

Consider a digital modulator output

$$\sum_{k=-\infty}^{\infty} z[k]p(t - kT), \tag{2.41}$$

with a real information-carrying sequence $z[k]$.

After channel propagation and receiver filtering with impulse responses $h_c(t)$ and $h_r(t)$ respectively, and additive noise $v(t)$, the received equivalent continuous-time signal is

$$r(t) = y(t) + v(t) = \sum_{k=-\infty}^{\infty} z[k]\, p_e(t - kT) + v(t) \qquad (2.42)$$

where $p_e(t) = (h_r * h_c * p)(t)$.

At the receiver, samples of $r(t)$ at time instants $t_n = nT + n_0T + \tau_n$ are obtained, where n_0T accounts for a global delay ($p_e(t)$ is centered around n_0T), and τ_n is a random variable that models the fluctuations in the local oscillator.

The resulting sequence is

$$r[n] = y[n] + v[n] =$$

$$\sum_{k=-\infty}^{\infty} z[k]p_e\left((n + n_0 - k)T + \tau_n\right) + v(t_n) =$$

$$\sum_{k=-\infty}^{\infty} z[k]q\left((n - k)T + \tau_n\right) + v(t_n) \qquad (2.43)$$

with the equivalent pulse $q(t) = p_e(t + n_0t)$ centered around the origin.

Focusing for now only on the signal part we can write

$$y[n] = \sum_{l=-\infty}^{\infty} z[n - l]h_n[l] \qquad (2.44)$$

with $h_n[l] = q(lT + \tau_n)$ showing the time-varying nature of the total response due to random sampling.

Equation (2.44) shows that each output value $y[n]$ is the result of a different filtering effect that depends on τ_n.

We are interested in understanding how the behavior of the stochastic process τ_n affects the signal $y[n]$.

To isolate the effects of the jitter from the time-invariant distortions of channel and receiver, we assume that the communication chain has been perfectly equalized in the absence of jitter. This is a reasonable assumption if τ_n is zero-mean and small when compared to T. We assume that receiver design, or adaptive equalization, is performed successfully around $\tau_n = 0$, and our analysis concerns the analysis of what happens around perfect equalization when τ_n is not null. Recall that $p(t)$, $h_c(t)$ and $h_r(t)$ must be bandlimited to $B < 1/T$ to allow equalization.

Elimination of inter-symbol interference (ISI) in the communication chain for $\tau_n = 0$ can be achieved in two ways:

• Make the cascade of the analog filters before sampling non distorting, i.e. pulse $p(t)$ and receiver filter $h_r(t)$ are chosen such that $q(t)$ satisfies Nyquist condition: $q(0) = 1$, $q(lT) = 0$, $l \neq 0$;
• Equalize via digital filters after sampling.

In the first case the effect of jitter, when $\tau_n \neq 0$, is a discrete time convolution with $h_n[l] = q(lT + \tau_n)$ composed of random samples of Nyquist's pulse $q(t)$.

In the second case equalization is performed on the discrete time sequence $r[n]$ after sampling and the equivalent pulse $q(t)$ does not satisfy Nyquist's condition. However, because of the bandlimited assumption, there must exist a Nyquist interpolator $i(t)$ such that

$$q(t) = \sum_{\alpha=-\infty}^{\infty} q(\alpha T) i(t - \alpha T). \tag{2.45}$$

Therefore, if in the absence of jitter for $\tau_n = 0$ equalization is achieved with a discrete-time impulse response $h_e[l]$

$$\sum_{l=-\infty}^{\infty} q(lT) h_e[m - l] = \delta[m], \tag{2.46}$$

in the presence of jitter we have

$$\sum_{l=-\infty}^{\infty} q(lT + \tau_n) h_e[m - l] =$$

$$\sum_{l=-\infty}^{\infty} \sum_{\alpha=-\infty}^{\infty} q(\alpha T) i(lT + \tau_n - \alpha T) h_e[m - l] = \tag{2.47}$$

$$\sum_{\beta=-\infty}^{\infty} i(\beta T + \tau_n) \sum_{l=-\infty}^{\infty} q\left((l - \beta)T\right) h_e[m - l] = i(mT + \tau_m).$$

The equivalent time-varying filter $h_n[l] = i(lT + \tau_n)$ is made up of samples of the Nyquist pulse.

Therefore, assuming that equalization has been achieved for $\tau_n = 0$, the typical effects of jitter must be analyzed in reference to the sampling patterns of the Nyquist interpolator.

Figure 2.10 shows a typical Nyquist pulse $i(t)$ and its sampling pattern for $\tau_n > 0$ and $\tau_n < 0$. The stochastic process τ_n is unknown and varies typically around zero.

Now the question is: is there any general property of the Nyquist pulse that can help in understanding the nature of the effects of the jitter in any system? A careful examination of the pulse shown in Fig. 2.10 reveals that

$$\sum_{l=-\infty}^{\infty} i(lT + \tau_n) = 1, \quad \forall \tau_n. \tag{2.48}$$

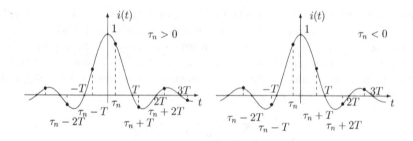

Fig. 2.10 Typical equivalent pulse and its sampling pattern for $\tau_n > 0$ and $\tau_n < 0$

Denoting with $I(f)$ the Fourier transform of $i(t)$, in the frequency domain (2.48) becomes

$$I(0) = T; \quad I\left(\frac{k}{T}\right) = 0, \ \forall \, k \neq 0. \tag{2.49}$$

Equation (2.49) is dual to the standard Nyquist condition $\sum_k I(f - k/T) = T$ applied to time and frequency domains in reversed order.

The equivalence between (2.48) and (2.49), can be immediately shown by observing that

$$\sum_{l=-\infty}^{\infty} i(lT + \tau_n) =$$

$$\int_{-\infty}^{\infty} e^{j2\pi f \tau_n} I(f) \sum_{l=-\infty}^{\infty} e^{j2\pi f lT} \, df =$$

$$\int_{-\infty}^{\infty} e^{j2\pi f \tau_n} I(f) \frac{1}{T} \sum_{k=-\infty}^{\infty} \delta\left(f - \frac{k}{T}\right) df =$$

$$\frac{1}{T} \sum_{k=-\infty}^{\infty} I\left(\frac{k}{T}\right) e^{j2\pi \frac{k}{T} \tau_n}.$$

Property (2.48) is rarely discussed in the literature and it is common to most reconstructing pulses.

It is worth to emphasize that such a condition is crucial *for any interpolation*, because when a pulse $i(t)$ is used to reconstruct an analog signal, we would like to ensure that from a constant sequence $x[n] = c$ after interpolation we get a constant signal $x(t) = c$, $\forall \, t$.

Most reconstructing pulses satisfy (2.49) also because all the pulses band-limited to $1/T$ are included. Examples are the sinc function $i(t) = sinc(t/T)$, the triangular function $i(t) = \Lambda(t/T)$ (piecewise linear interpolation) and the raised

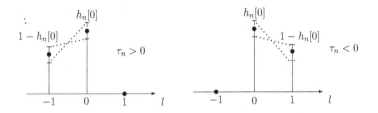

Fig. 2.11 Simplified equivalent discrete-time impulse response with complementary samples for $\tau_n > 0$ and $\tau_n < 0$

cosine function $i(t) = \frac{\cos \pi r \frac{t}{T}}{1 - 4r^2 t^2} sinc(t/T)$, where $0 \leq r \leq 1$ is the roll-off factor corresponding to bandwidth $\frac{1}{2T} \leq W \leq \frac{1}{T}$.

In modeling the timing jitter process, we observe that τ_n causes a time-varying filtering effect with an impulse response $h_n[l]$ with values that sum up to one for each τ_n. To simplify the analysis, since reconstruction pulses are always mostly concentrated on the main lobe and τ_n is small, we can assume that the relevant samples are only two as shown in Fig. 2.11.

In practice, the proposed simplified model assumes that, after jitter, each sequence element at time n is the result of a convex linear combination of the current sample and one of the two adjacent samples. Even the simplicity of this three-pulse model captures the essence of the jitter's effects, confirming results of other studies on the topic that are indicated in the references at the end of this chapter. In the normalized frequency domain for a fixed τ_n we have

$$H_n(\nu) = \begin{cases} h_n[0] + (1 - h_n[0])e^{j2\pi\nu} & \tau_n > 0 \\ h_n[0] + (1 - h_n[0])e^{-j2\pi\nu} & \tau_n < 0 \end{cases}$$

with energy transfer function

$$|H_n(\nu)|^2 = 1 - 2(1 - \cos 2\pi\nu)h_n[0](1 - h_n[0]) \tag{2.50}$$

that does not depend on the sign of τ_n.

Since τ_n is random, also $|H_n(f)|^2$ is random and depends on the statistics of $h_n[0]$. The process τ_n can be modeled as a stationary process that is the superposition of a constant misalignment τ_0 and a random zero-mean stationary process e_n, $\tau_n = \tau_0 + e_n$, even though most of the time we assume $\tau_0 = 0$. The shape of the specific pulse $i(t)$ and the pdf of τ_n determine the statistics of $h_n[0]$

$$E[h_n[0]] = \overline{h};$$
$$E[h_n^2[0]] = m_2;$$

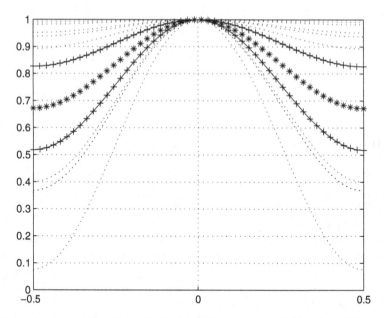

Fig. 2.12 Family of $|H_n(\nu)|^2$ for various realizations of $h_n[0]$ coming from Gaussian samples of τ_n with $\sigma_t = 0.3T$, and $i(t) = sinc(t/T)$. The curve with the asterisks (*) is the mean $E\left[|H_n(\nu)|^2\right]$ and the other two (+) are $E\left[|H_n(\nu)|^2\right] \pm \frac{1}{2}Std\left[|H_n(\nu)|^2\right]$

$$VAR[h_n[0]] = \sigma_h^2;$$

$$E[h_n^3[0]] = m_3;$$

$$E[h_n^4[0]] = m_4. \tag{2.51}$$

Mean and standard deviation of the energy transfer function are

$$E\left[|H_n(\nu)|^2\right] = 1 - 2(1 - \cos 2\pi\nu)\left[\overline{h}(1 - \overline{h}) - \sigma_h^2\right], \tag{2.52}$$

$$Std\left[|H_n(\nu)|^2\right] = 2(1 - \cos 2\pi\nu)\sqrt{m_2 + m_4 - 2m_3 - \overline{h}^2 - m_2^2 + 2\overline{h}m_2}. \tag{2.53}$$

Figure 2.12 shows a collection of plots of $|H_n(\nu)|^2$ for various values of $h_n[0]$ obtained from samples of τ_n with $\tau_0 = 0$ and e_n gaussian with $\sigma_e = 0.3\,T$. It has a low-pass effect and the variability, as predicted, is larger at the highest frequencies. The mean behavior is also shown with the confidence interval $(\pm\frac{Std}{2})$ that increases as ν approaches the highest frequency $\nu = 0.5$. The deleterious effects described by the standard deviation, that grows at the higher frequencies, confirm the results reported in the literature about the effects of jitter in OFDM systems where the channels corresponding to the largest frequencies, are most unstable. The OFDM system is the optimum candidate for showing the frequency selective effect of the timing jitter.

2.3.2 Jitter in OFDM

In this section the authors analyze more specifically the effects of our jitter model on the performance of a simplified OFDM modulation system. We skip over the standard details of the OFDM chain and we assume that jitter, after the introduction of the cyclic prefix, appears in the circular convolution as the time-varying filter response derived in the previous section.

To isolate jitter's effects we assume, as in the previous section, that the channel does not introduce distortion on the signal. We would like to remind the reader that this does not remove generality from our analysis because any time-invariant distortion can be eliminated in the OFDM system by scaling and rotating each channel. Jitter, instead, is unpredictable and it is analyzed in its unavoidable effects around perfect equalization conditions.

The analysis carried out here emphasizes the specific role of jitter on a finite-length window, while the results at the end of the previous section can be considered as a long-term analysis for any sampled system.

In a typical OFDM chain, at each symbol time k, the source input $a[k]$ is a vector of N complex symbols associated with the various frequency channels. Each component of $a[k]$ takes values on a two-dimensional constellation. An $N \times N$ inverse discrete Fourier transform (IDFT) is computed by a block

$$W^H = \frac{1}{\sqrt{N}}[e^{j\frac{2\pi}{N}mn}]_{mn}, \ m, n = 0, ..., N - 1 \qquad (2.54)$$

as $z[k] = W^H a[k]$.

Skipping over the details about the cyclic prefix, the noisy block is received (with no channel distortion) as

$$\mathbf{r}[k] = H[k]W^H \mathbf{a}[k] + H[k]\mathbf{v}[k] \qquad (2.55)$$

where $H[k]$ models jitter's effects, and $\mathbf{v}[k]$ is a vector of independent noise with covariance matrix $\sigma_v^2 I_N$.

The final system output is

$$\mathbf{y}[k] = WH[k]W^H \mathbf{a}[k] + WH[k]\mathbf{v}[k]. \qquad (2.56)$$

The matrix $H[k]$ has the structure

$$
H[k] = \begin{pmatrix}
h_0^k[0] & h_0^k[1] & 0 & 0 & 0 & 0 & 0 & . & h_0^k[-1] \\
h_1^k[-1] & h_1^k[0] & h_1^k[1] & 0 & 0 & 0 & 0 & . & 0 \\
0 & h_2^k[-1] & h_2^k[0] & h_2^k[1] & 0 & 0 & 0 & . & 0 \\
& . & . & . & . & . & . & & . \\
h_{N-1}^k[1] & 0 & 0 & 0 & 0 & 0 & 0 & h_{N-1}^k[-1] & h_{N-1}^k[0]
\end{pmatrix}
$$

where, in general, at each frame k, and at each row, there is a different impulse response.

The vector $\left(h_n^k[-1], h_n^k[0], h_n^k[1] \right)$ has been wrapped around because of the cyclic prefix. Strictly speaking this is just a simplified equivalent model also because the cyclic prefix should have no circular effects on the noise term. For simplicity we prefer to avoid lengthy discussions on these details also because the differences are really minor and confined to the upper-right and lower-left corners of the matrix. They are practically irrelevant for realistic values of N.

The impulse response is chosen to be real assuming that the two simultaneous channels (real and imaginary part) are sampled at the same time (although the analysis can be easily generalized). In each frame there is a set of sampling time delays $\left(\tau_0^k, \tau_1^k, ..., \tau_{N-1}^k \right)$ that corresponds to $\left(h_0^k[0], h_1^k[0], ..., h_{N-1}^k[0] \right)$.

Straightforward matrix calculations give

$$
C[k] = W H[k] W^H =
$$

$$
\frac{1}{N} \left[\sum_{q=0}^{N-1} e^{-j\frac{2\pi}{N}q(n-m)} \left[h_q^k[0] + \left(1 - h_q^k[0]\right) e^{j \, \mathrm{sgn}(\tau_q^k)\frac{2\pi}{N}m} \right] \right]_{n,m=0,...,N-1} .
$$

$$
(2.57)
$$

Since $C[k]$ is in general non-diagonal we have that *jitter causes temporary loss of orthogonality and unequal effects on the various channels.* Also the noise term is affected with a weak time-varying coloring effect.

Stationarity implies that

$$
E[h_n[0]] = \overline{h} \quad \forall \, n \tag{2.58}
$$

and

$$
E\left[\left(h_n^k[-1], h_n^k[0], h_n^k[1] \right) \right] = \left(\frac{1-\overline{h}}{2}, \overline{h}, \frac{1-\overline{h}}{2} \right) . \tag{2.59}
$$

The output mean-behavior is preserved, because for every source vector $a[k]$, we have

$$
E\left[y[k] | a[k] \right] = W \overline{H} W^H a[k] = diag\left(\lambda_0, ..., \lambda_{N-1} \right) a[k] \tag{2.60}
$$

where

$$\lambda_m = \overline{h} + (1 - \overline{h}) \cos \frac{2\pi}{N} m, \ m = 0, ..., N - 1 \qquad (2.61)$$

and $\overline{H} = E[H[k]]$ is Toeplitz and circulant.

Therefore, on the average, the constellations are scaled non uniformly with larger attenuations for the channels corresponding to the highest frequencies (remember that the middle channels are the ones corresponding to the highest frequencies). This average low-pass effect can be easily neutralized via proper channel power control, but performance will be unavoidably affected because of the variations around the mean. Such instabilities are described by the conditional covariance matrix

$$Cov\left[\mathbf{y}[k] | \mathbf{a}[k]\right] =$$

$$WE\left[\left(H[k] - \overline{H}\right) W^H \mathbf{a}[k]\mathbf{a}[k]^H W \left(H[k] - \overline{H}\right)^T [3pt]\right] W^H +$$

$$\sigma_v^2 WE\left[H[k]H^T[k]\right] W^H \qquad (2.62)$$

that contains a signal dependent term and a noise term.

In general the signal dependent term is different for every complex input vector

$$\mathbf{a}[k] = (\alpha_0[k] + j\beta_0[k], ..., \alpha_{N-1}[k] + j\beta_{N-1}[k]) \qquad (2.63)$$

also because each component may use a different constellation (non uniform channel loading). Note that also the noise covariance is affected by the time-varying nature of $H[k]$.

To obtain here a compact description of jitter's effects, we assume for simplicity that all the constellations are defined on the same squared $M \times M$ lattice and that they are uniformly and independently associated to the incoming bitstream. Therefore

$$E[\mathbf{a}[k]\mathbf{a}[k]^H] = (\sigma_\alpha^2 + \sigma_\beta^2)I_N = \sigma_a^2 I_N. \qquad (2.64)$$

Averaging over all the possible constellations we get the global covariance

$$Cov[\mathbf{y}[k]] =$$

$$\sigma_a^2 W \left(E\left[H[k]H[k]^T\right] - \overline{H}\,\overline{H}^T\right) W^H + \sigma_v^2 WE\left[H[k]H[k]^T\right] W^H \qquad (2.65)$$

with the matrix $E\left[H[k]H[k]^T\right]$ having the structure

$$E\left[H[k]H[k]^T\right] =$$

$$E \begin{pmatrix} x & x & x\ 0\ 0\ 0\ .\ 0\ h_0[-1]h_{N-2}[1] & \sum_{i=-1}^{0} h_0[i]h_{N-1}[i+1] \\ x & x & x\ x\ 0\ 0\ .\ 0 & 0 & h_1[-1]h_{N-1}[1] \\ . & . & . \ . \ . \ . \ . \ . & . & . \\ 0 & 0 & x\ x\ x\ x\ x\ 0 & . & 0 \\ . & .. & . \ . \ . \ . \ . & . & . \\ h_{N-2}[1]h_0[-1] & 0 & 0\ 0\ 0\ x\ x & x & x \\ \sum_{i=-1}^{0} h_{N-1}[i+1]h_0[i] & h_{N-1}[1]h_1[-1]\ 0\ 0\ 0\ 0\ x & x & x \end{pmatrix}$$

with generic row

$$\left(0,.,0, h_n[-1]h_{n-2}[1], \ \overset{0}{\underset{i=-1}{\sum}} h_n[i]h_{n-1}[i+1], \ \overset{1}{\underset{i=-1}{\sum}} h_n^2[i], \right.$$

$$\left. \overset{1}{\underset{i=0}{\sum}} h_n[i]h_{n+1}[i-1], \ h_n[1]h_{n+2}[-1], \ 0,.,0\right)$$

where we have dropped the superscript k for notation simplicity.

The correlations involved in the matrix are computed from knowledge of the autocorrelation of the stochastic process $\{h_n[0]\}$ denoted $r_l = E[h_n[0]h_{n-l}[0]]$, that can be computed from knowledge of the autocorrelation of the gaussian process $\{\tau_n\}$. More specifically, the time characterization of $\{\tau_n\}$ determines the time evolution of the jitter filter parameter $\{h_n[0]\}$, that ultimately determines the variations on the various channels.

Straightforward calculations carried out by conditioning on the sign of τ_n, give the moments

$$E\left[h_n^2[0]\right] = r_0;$$

$$E\left[h_n^2[1]\right] = E\left[h_n^2[-1]\right] = \frac{1}{2}(1 - 2\overline{h} + r_0);$$

$$E\left[h_n[0]h_{n+1}[-1]\right] = E\left[h_n[1]h_{n+1}[0]\right] = \frac{1}{2}(\overline{h} - r_1);$$

$$E\left[h_n[1]h_{n+2}[-1]\right] = E\left[h_n[-1]h_{n-2}[1]\right] = \frac{1}{4}(1 - 2\overline{h} + r_2);$$

$$E\left[h_0[-1]h_{N-2}[1]\right] = E\left[h_1[-1]h_{N-1}[1]\right] = \frac{1}{4}(1 - 2\overline{h} + r_{N-2});$$

$$E\left[h_0[0]h_{N-1}[1]\right] = E\left[h_0[-1]h_{N-1}[0]\right] = \frac{1}{2}(\overline{h} - r_{N-1}). \qquad (2.66)$$

Matrix $\overline{H}\,\overline{H}^T$ has the Toeplitz structure

$$\overline{H}\,\overline{H}^T = E \begin{pmatrix} x & x & x & 0 & 0 & 0 & . & 0 & \left(\frac{1-\overline{h}}{2}\right)^2 & \overline{h}\left(1-\overline{h}\right) \\ x & x & x & x & 0 & 0 & . & 0 & 0 & \left(\frac{1-\overline{h}}{2}\right)^2 \\ . & . & . & . & . & . & . & . & & . \\ 0 & 0 & x & x & x & x & x & 0 & . & 0 \\ . & .. & & . & . & . & . & . & & . \\ \left(\frac{1-\overline{h}}{2}\right)^2 & 0 & 0 & 0 & 0 & x & x & & x & x \\ \overline{h}(1-\overline{h}) & \left(\frac{1-\overline{h}}{2}\right)^2 & 0 & 0 & 0 & 0 & x & & x & x \end{pmatrix}$$

with the generic row

$$\left(0,.,0, \left(\frac{1-\overline{h}}{2}\right)^2, \overline{h}(1-\overline{h}), \frac{3}{2}\overline{h}^2 - \overline{h} + \frac{1}{2}, \overline{h}(1-\overline{h}), \left(\frac{1-\overline{h}}{2}\right)^2, 0,.,0\right).$$

Putting everything together we have

$$E\left[H[k]H[k]^T - \overline{H}\,\overline{H}^T\right] =$$

$$E \begin{pmatrix} x & x & x & 0 & 0 & 0 & . & 0 & \frac{1}{4}(r_{N-2} - \overline{h}^2) & -(r_{N-1} + \overline{h}^2) \\ x & x & x & x & 0 & 0 & . & 0 & 0 & \frac{1}{4}(r_{N-2} - \overline{h}^2) \\ . & . & & . & . & . & . & . & & . \\ 0 & 0 & x & x & x & x & x & 0 & . & 0 \\ . & .. & & . & . & . & . & . & & . \\ \frac{1}{4}(r_{N-2} - \overline{h}^2) & 0 & 0 & 0 & 0 & x & x & & x & x \\ -(r_{N-1} + \overline{h}^2) & \frac{1}{4}\left(r_{N-2} - \overline{h}^2\right) & 0 & 0 & 0 & 0 & x & & x & x \end{pmatrix}$$

with generic row

$$\left(0,.,0,\frac{1}{4}(r_2 - \overline{h}^2), \ -(r_1 - \overline{h}^2), 2r_0 - \frac{3}{2}\overline{h}^2 - \overline{h} + \frac{1}{2},\right.$$

$$\left.-(r_1 - \overline{h}^2), \frac{1}{4}(r_2 - \overline{h}^2), 0,.,0\right).$$

Matrix $E\left[H[k]H[k]^T - \overline{H}\,\overline{H}^T\right]$ is not exactly circulant because of the terms around the upper-right and lower-left corners. However, if N is sufficiently large, we can assume with reasonable approximation that

$$W E\left[H[k]H[k]^T - \overline{H}\,\overline{H}^T\right] W^H \simeq diag\left(g_0, ..., g_{N-1}\right) \qquad (2.67)$$

with

$$g_m = 2r_0 - \frac{3}{2}\overline{h}^2 - \overline{h} + \frac{1}{2} - 2(r_1 - \overline{h}^2)\cos\frac{2\pi}{N}m + \frac{1}{2}(r_2 - \overline{h}^2)\cos\frac{2\pi}{N}2m,$$

$$m = 0, ..., N - 1. \tag{2.68}$$

The channels are essentially uncorrelated, but their variance becomes in general unequally distributed.

Similarly for the noise term we have

$$WE\left[H[k]H[k]^T\right]W^H \simeq diag\left(d_0, ..., d_{N-1}\right) \tag{2.69}$$

with

$$d_m = 2r_0 + 1 - 2\overline{h} - 2(r_1 - \overline{h})\cos\frac{2\pi}{N}m + \frac{1}{2}(r_2 - 2\overline{h} + 1)\cos\frac{2\pi}{N}2m,$$

$$m = 0, ..., N - 1. \tag{2.70}$$

The noise components are essentially uncorrelated, but there is, in general, an uneven effect on the various channels.

To look more in the details of jitter's effects on OFDM performance, we distinguish here two scenarios:

- *Independent jitter*: at each time-sample the local oscillator is shifted by values $\{\tau_n\}$ that are independent and identically distributed (i.i.d.);
- *Correlated jitter*: the time series $\{\tau_n\}$ exhibits a time-correlation. This second case can model slower jitter, with subsequent delay values anchored to the previous ones.

Independent jitter: the process $\{\tau_n\}$ is gaussian zero-mean with autocorrelation $R_\tau[l] = \sigma_\tau^2\delta[l]$. Therefore also the auto-covariance of $h_n[0]$ is i.i.d. with

$$r_l - \overline{h}^2 = (r_0 - \overline{h}^2)\delta[l]. \tag{2.71}$$

At the output, all the channels are essentially uncorrelated and they have practically the same variance. The extra variance term caused by jitter can be summed up to the noise variance, that remains essentially white. The total value to be used for computation of the error probability is

$$\sigma^2 = \sigma_a^2\left(2r_0 - \frac{3}{2}\overline{h}^2 - \overline{h} + \frac{1}{2}\right) + \sigma_v^2\left(2r_0 + 1 - 2\overline{h}\right). \tag{2.72}$$

To verify that, note that in the absence of jitter, $\tau_n = 0$, $h_n[0] = 1$, $\overline{h} = 1$, $r_0 = E[h_n^2[0]] = 1$ and $\sigma^2 = \sigma_v^2$.

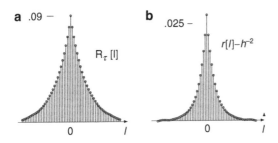

Fig. 2.13 (**a**) Auto-correlation $R_\tau[l]$ of the Gaussian jitter process $\{\tau_n\}$ for $\sigma_\epsilon = 0.13$ and $\alpha = 0.9$; (**b**) auto-covariance $r[l] - \overline{h}^2$ of $\{h_n[0]\}$ for the *sinc* interpolator

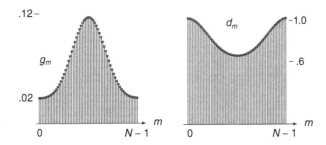

Fig. 2.14 The variance distribution as a function of the carrier index m for the signal term g_m and the noise term d_m

Correlated jitter: within each OFDM window, delay values may vary around zero at a speed that depends on the ratio between symbol rate and sampling frequency. As a first-order modeling of the time evolution of τ_n we assume that $\tau_n = \alpha\tau_{n-1} + \epsilon_n$, with ϵ_n white sequence with variance σ_ϵ^2. The autocorrelation is $R_\tau[l] = \frac{\sigma_\epsilon^2}{1-\alpha^2}\alpha^{|l|}$. The mean and the autocorrelation of $\{h_n[0]\}$, that depends on the chosen interpolator, are in principle obtainable in analytical closed-form. The calculations are a bit cumbersome and for simplicity, we prefer to compute them numerically.

Figure 2.13 shows the autocorrelation $R_\tau[l]$ and the auto-covariance $r[l] - \overline{h}^2$ for a gaussian sequence τ_n with $\sigma_\epsilon = 0.13$, $\alpha = 0.9$, and the *sinc* interpolator. The variances g_m and d_m for $N = 64$ are plotted in Fig. 2.14 as function of the carrier index m.

Note that the normalized frequencies are $\nu \in \{0, \frac{1}{N}, ..., \frac{N-1}{N}\}$ and the central values are those corresponding to the largest frequencies.

It is clear from this example, that the highest frequencies in the signal term are the most affected by the timing jitter while on the noise we find a weaker low-pass effect.

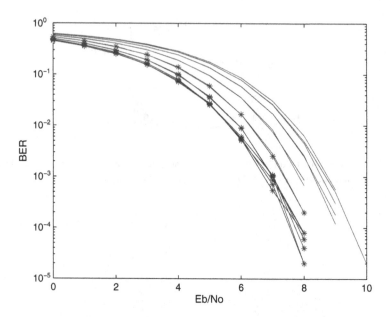

Fig. 2.15 Performance of a simulated OFDM chain with correlated timing jitter. The curves with the asterisk correspond to carrier indexes $0, 1, 2, 3, 12, 13, 14, 15$ (lower frequencies) while the others correspond to $4, 5, 6, 7, 8, 9, 10, 11$ (higher frequencies)

2.3.3 Simulation Results

In this section we show the system performance in terms of symbol error rate. They can be computed on the scaled signals $\sqrt{\lambda_m} a_m$, $m = 0, ..., N - 1$ and variance distribution $\sigma_m^2 = \sigma_a^2 g_m + \sigma_v^2 d_m$, $m = 0, ..., N - 1$. Clearly jitter's effects are more marked when the signal-to-noise ratio is high. Also, a minimum distance detector, because of slightly correlated noise, is slightly sub-optimal.

Figure 2.15 shows the bit error rate for a simulated OFDM system with $N = 16$ sub-carriers all loaded uniformly with $M = 4 \times 4$ symbols. Jitter's parameters are $\alpha = 0.9$, $\sigma_\epsilon = 0.13$ and the receiver is multiple-threshold detector. Clearly the carriers corresponding to the highest frequencies are affected the most.

2.4 Two Channels Time Interleaved Analog to Digital Converters

A common way to overcome the limits of the hardware technology that affects the ADC sampling frequency is to time interleave two, or more, analog to digital converters. In this section of the book the authors refer to a two channels (or paths) time interleaved ADC system. The goal is to cancel in the sample data domain the artifacts derived from the analog hardware devices (ADCs).

2.4.1   Time Interleaved ADCs: The Problem

By using two ADCs operating in parallel with a T_s time offset of their $2T_s$ time interval sampling clocks, the overall sampling frequency of the system is doubled. In an ideal two-channel TI-ADC, the aliasing terms formed by the individual ADCs, operating at half rate, are canceled by the interleaving process. This cancelation occurs because the aliased spectral component of the time offset ADC has the opposite phase of the same spectral component of the non-time offset ADC. In the absence of time offset and gain mismatch the sum of their spectra would cancel their alias components.

Because of gain and timing phase mismatches of the analog hardware components in the ADCs, the spectral aliasing components from the interleaved time series replicas do not sum to zero. The sampling instants of the two ADCs are, in fact, affected by a constant delay, Δt_m with $m = 0, 1$, which results in an undesired frequency dependent phase offset of their aliased spectra that prevents their perfect cancelation at the output of the time multiplexer.

Mismatches in path gains g_m, with $m = 0, 1$, of the TI-ADC, due to tolerance spread of analog components are always present in the ADC's hardware. The gain mismatch contributes an imperfect frequency dependent cancelation of the spectral components at the output of the TI-ADC system. In order to correct the artifacts caused by the time and gain offsets, of course, we must first estimate them.

Estimation methods can be divided in two categories:

- Foreground techniques, also known as non-blind, that inject a known test or probe signal to estimate the mismatches by measuring the TI-ADC output responses to the probe.
- Background techniques, also known as blind, for which no information is required about the input signal (except perhaps for some knowledge about the presence or absence of signal activity in certain frequency bands) in order to estimate the mismatches.

The first approach has the disadvantage that normal TI-ADC operations are suspended during the probe while in the second approach the calibration process does not interrupt normal operation.

Many works present in the literature are regarding blind (and non-blind) estimation and correction methods for both, time and gain offsets, in a two channels TI-ADC; many authors have also extended well-known results to the M channels TI-ADC case. Because all the estimation and correction structures are derived assuming the low-pass nature of the input signal, they do not work if applied to intermediate frequency signals, i.e. communication scenario.

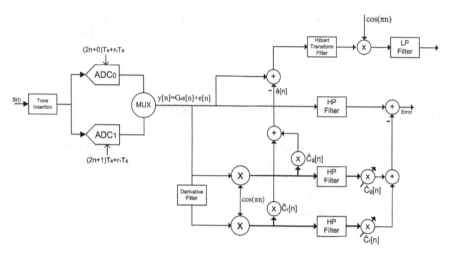

Fig. 2.16 Block scheme of the proposed structure

In this book the authors derive a semi-blind approach to adaptively estimate and correct both time and gain offsets in a two channel TI-ADC operating on communication signals. The injection of a low level probe tone signal between DC and the low frequency end of the signal's band centered spectrum allows the application of well known blind estimation and correction schemes. The knowledge of the probe's frequency allows us to easily remove it from the base-band down converted signal once it has been recovered. A Hilbert transform filter followed by a digital down converter represents the best solution for shifting to base-band the signal after mismatch corrections.

2.4.2 Proposed Solution

The block scheme of the proposed structure for estimation and compensation, in the sample data domain, of both, gain and time mismatches occurring in a two channel TI-ADC, is shown in Fig. 2.16.

It is capable of operating on band-pass signals. Its core is based on the structures derived in the literature, respectively for correcting time mismatches in a two channel TI-ADC and time and gain mismatches in an M-channel TI-ADC (see references at the end of the chapter).

These structures have been derived by modeling the time and gain offsets of the ADCs as in Fig. 2.16 and by identifying, at the output of TI-ADC, the different contributions of the signal and of the error caused by the mismatches.

The error estimate is based on the assumptions that the timing offsets are small relative to the overall sampling period, and their average value is zero. The basic observation that underlies both of these structures is that by slightly oversampling the input signal $s(t)$ with a TI-ADC in which no mismatches occur, we should be able to observe some spectral regions where no signal energy is present. However, because of gain and time offsets between the two channels, a certain amount of undesired energy appears in these bands (called mismatching bandwidths). By filtering and minimizing this energy it is possible to adaptively identify and correct both of the mismatches and, for this purpose the least-mean-square (LMS) algorithm is a natural option.

Note that the assumption on the low-pass nature of the input signal, along with the knowledge of the sampling frequency, allows us to predict the position of the mismatching bandwidths and to design high-pass filters to isolate, monitor, and correctly process these spectral regions.

In the case of a communication scenario, because of the band-pass nature of the involved signals, the main assumption, regarding the spectral position of the mismatching bandwidth, on which the above mentioned structures are based, cannot be applied.

In a common digital receiver, with only one analog to digital converter, the sampling frequency f_{ST} is selected in order to satisfy the equality $f_{ST} = 2f_{MAX} + \Delta f$ where Δf is called oversampling factor. Note that it represents the gap between the two signal spectra replicas after sampling.

It is our interest to choose this factor as small as possible, compatible with the requirements for the subsequent filtering tasks that the digital receiver has to handle.

In a practical receiver usually

$$0 < \Delta f \leq 2f_{IF} - BW. \tag{2.73}$$

In the most common case of equality $\frac{f_{IF}}{f_{ST}} = \frac{1}{4}$.

Note that the case $\Delta f \geq 2f_{IF} - BW$ corresponds to a gap between f_{MAX} and $\frac{f_{ST}}{2}$ that is bigger than the gap between the zero frequency and f_{MIN} in the first Nyquist zone. This hypothesis is commonly discarded because it implies wastage of bandwidth.

When a two channel TI-ADC is used, the same sampling frequency $f_{s_{0,1}} = \frac{f_{ST}}{2}$ is used on each arm with a time shift of the initial sampling time in one of the arms equal to $\frac{1}{f_{ST}}$. These sampling frequencies violate Nyquist sampling theorem and, as a consequence, the negative side of the replica that resides in the second Nyquist zone appears in the first Nyquist zone. This replica should be automatically suppressed at the output of the multiplexer if no mismatches are present in the structure.

We specified before that $\frac{\Delta f}{2}$ represents the gap between f_{MAX}, the maximum frequency of the input signal, and $f_{s_{0,1}}$. It also represents the gap between the zero frequency and the minimum frequency of the negative replica coming from the second Nyquist zone.

In the case in which

$$\frac{\Delta f}{2} = f_{IF} - \frac{BW}{2} \tag{2.74}$$

the two replicas, the positive one belonging to the first Nyquist zone and the negative one belonging to the second Nyquist zone, will perfectly overlap on each other and it is difficult to visualize the mismatches caused by the time and gain offset.

In the case in which

$$\frac{\Delta f}{2} \leq f_{IF} - \frac{BW}{2} \tag{2.75}$$

the negative side of the signal replicas belonging to the second Nyquist zone will partially overlap on the positive signal part belonging to the first Nyquist zone; it will be, in fact, closer to zero.

It is clear that, for both of the cases specified above, we will not have undesired energy between f_{MAX} and $f_{s0,1}$. This is exactly the spectral region in which the mismatching bandwidths have been defined and for which the high-pass filters should be designed.

In order to present a certain amount of energy to the high-pass filters for allowing the estimation of the multiplier coefficients $\hat{c}_g[n]$ and $\hat{c}_r[n]$ (see Fig. 2.16) that are adapted to minimize this energy in the mean-square sense, we insert a carrier signal (or pilot tone) $sin(2\pi f_c t)$ in a proper position, between DC and f_{MIN}, in the analog spectrum of the TI-ADC input signal.

Note that f_c should not to be too close to DC, because that will cause an undesired increase in the convergence time of the LMS algorithm, on the other side, on each arm, after sampling we would like f_c to be as far as possible from the information signal, so that we can use a smaller number of taps in the design of the high-pass filters for the identification task. From the previous derivations, it is clear that $f_c \in (0, \frac{\Delta f}{2})$ and, as a consequence of the above cited reasons, here, we use $f_c = \frac{\Delta f}{4}$. Note that, as a consequence of the spectral folding caused by Nyquist sampling theorem violation occurring in each arm, the sine wave falls exactly in the passband of the high-pass filters.

Once the error estimation and correction is done, of course the receiver has to down convert the input signal to base-band and filter it in order to remove the residual sine wave that was previously inserted. By knowing its exact position this will not be a difficult task. Before shifting the signal to DC, it is processed by a Hilbert transform filter, this is surely an inexpensive and easy solution that has the double effect of canceling the negative part of the input signal and down sampling it 2-to-1. If the overall sampling frequency is selected to satisfy the equality in (2.73), after being processed by the Hilbert transform, the complex signal spectrum resides around 0.5 in the normalized frequency domain with its sampling frequency halved; at this point, a digital down sampler, that is actually a sequence of $(-1)^n$ multiplied with the signal, shifts it to base-band where it can easily be low-pass filtered for removing the probing sine wave that was inserted in the analog domain.

Note that along with the probing tone, as consequence of this last filtering process, also the DC offset is eliminated.

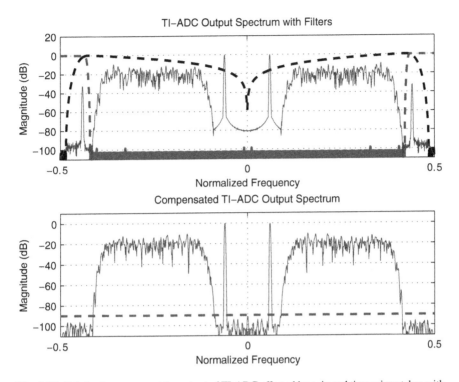

Fig. 2.17 Subplot 1: spectrum at the output of TI-ADC affected by gain and time mismatches with superimposed derivative and high-pass filters; subplot 2: spectrum at the output of the identification and correction structure

2.4.3   Simulation Results

In this section the authors present some simulations to show the effectiveness of the proposed structure for canceling time and gain mismatches in a two channel TI-ADC.

In the first subplot of Fig. 2.17 the spectrum of a QPSK signal processed by a two-channel TI-ADC with gain and time offsets is shown. The sine tone injected in the analog domain at the normalized frequency $f_c = 0.06$ is also shown. The timing offsets for this example are $r_0 = 0$ and $r_1 = 0.04$ which corresponds to a 4% error in the overall sampling time. Note that the first time delay is used as a reference point in the simulations. The gain offsets are $g_0 = 0$ and $g_0 = 0.05$ which correspond to a 5% error on the second arm of the TI-ADC. The spectra of the derivative and the high-pass filters designed for the errors estimation structure are superimposed in this plot. As Nyquist sampling criterion is violated, a spectral copy of the training tone appears at the normalized frequency of 0.44 that is exactly located in the mismatching bandwidth filtered by the high-pass filters. Note that the QPSK signal replica that should belong to the second Nyquist zone also appears

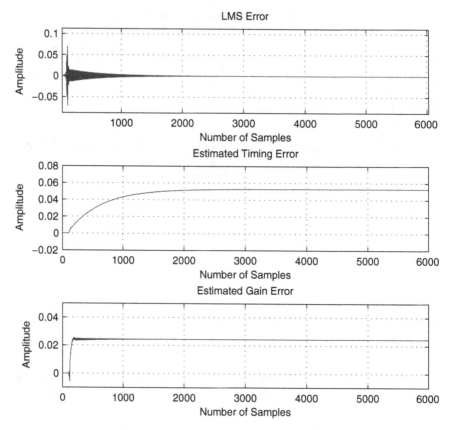

Fig. 2.18 Subplot 1: convergence behavior and estimated LMS error; subplot 2: estimated gain error; subplot 3: estimated timing error

in the first one. This replica is completely superimposed on the signal and, for this reason, it is not possible to demonstrate its presence. In the second subplot of Fig. 2.17 the signal spectrum at the output of the compensator is shown. The training tone due to mismatches is now absent: by using the proposed estimation and compensation structure we are able to reduce its energy below 90 dB. This value is indicated by the dotted line in the figure.

Figure 2.18 shows the convergence behavior of the estimation process. The LMS error converges when the timing and gain errors converge to their correct values. The chosen step for the LMS algorithm in this example is $\mu = 0.04$. Because the LMS algorithm has been applied to minimize the energy of a deterministic sine tone, the converged value of the error has zero mean with zero variance. In the second subplot of Fig. 2.18, the convergence behavior of the weight associated with the timing error is shown. It converges to the appropriate value after 4,000 samples. Similarly, the third subplot of Fig. 2.18 shows the convergence process of the weight associated

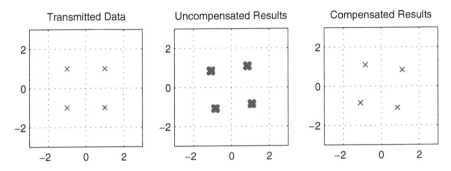

Fig. 2.19 Subplot 1: transmitted QPSK constellation; subplot 2: effect of gain and timing mismatches on the transmitted QPSK constellation; subplot 3: effect of the correction structure on the output QPSK constellation of the TI-ADC

with the gain error estimation; the process converges after 200 samples to 0.025 which is the theoretical expected value corresponding to the average value of the ADCs gains.

In order to demonstrate the degree of mismatch suppression which we cannot see directly in the spectral plots, we compare in Fig. 2.19 the demodulated QPSK constellation (after correction) with the transmitted one. The demodulation process is achieved by passing the signal through a Hilbert transform, which gives us access to the analytic signal and its complex envelope. The signal is then down converted by a complex heterodyne in a digital down converter. A further low-pass filter is applied to suppress the probe tone. Finally, the matched filter, with the proper time alignment, but no phase correction, is applied to the complex base-band signal to maximize its signal to noise ratio.

The constellation resulting from this process is shown in the third subplot of Fig. 2.19 along with the transmitted QPSK constellation in the first subplot and the corrupted QPSK constellation at the output of TI-ADC in the second subplot. It is clearly shown that the TI-ADC mismatches result in an increased variance cloud around the matched filter output constellation points. The variance clouds are completely removed by using the proposed structure.

We now apply the proposed suppression technique to a different alias distorted signal to better illustrate its performance. We note that it is a jump in faith to assume that suppressing the probe tone located in the out-of-band spectral region leads to similar suppression levels of spectral artifacts in the in-band spectral region. In Fig. 2.20, we show the results of the second example that completely demonstrates the effectiveness of the proposed structure. In this example, we generated 17 equally spaced sine waves spanning frequencies from 0.1 to 0.4 on the normalized frequency axis. The training tone is still located at $f_c = 0.06$ and the gain and time errors are the same as those used in the simulation of Fig. 2.17. The combined effects of time and gain offsets can be visualized in the first subplot of Fig. 2.20 where the folded spectrum, coming from the second Nyquist zone and unsuppressed at the output of the multiplexer, appear between the spectral lines of the constructed

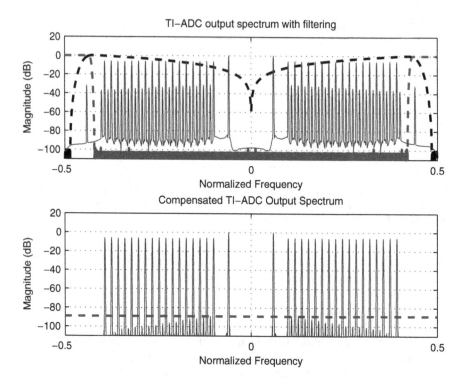

Fig. 2.20 Subplot 1: sine wave spectrum at the output of TI-ADC affected by gain and time mismatches with superimposed derivative and high-pass filters; subplot 2: sine wave spectrum at the output of the identification and correction structure

information signal. The second subplot of Fig. 2.20 shows the spectrum obtained after compensation. Here we can clearly recognize that the spectral artifacts are significantly reduced while the training tone is completely suppressed. We also note a residual spectrum containing the artifact remnants that were not suppressed to the same degree as the probe signal. It is below −90 dB that is underlined by means of the dotted line in the same picture.

Note that before compensation, the maximum amplitude, on log scale, of the spurious peaks affecting the signal is −30.2 dB; after compensation their maximum amplitude is −90 dB. This result clearly demonstrates that the structure proposed by the authors is capable of obtaining improvement of approximately 60 dB.

It is interesting to look at the residual mismatching spectrum inside the signal information bandwidth; it has a frequency dependent amplitude with a slope reminiscent of a frequency shifted derivative filter and, we believe that this is an uncompensated residual timing error likely attributed to higher order Taylor Series terms in the error approximation. This observation opens the door for further research in the direction of improving the proposed structure for complete suppression of the residual mismatch spectrum within the signal spectrum. For

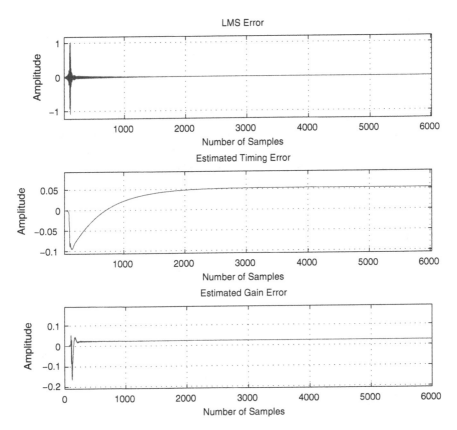

Fig. 2.21 Subplot 1: convergence behavior and estimated LMS error for the sine wave spectrum; subplot 2: estimated gain error; subplot 3: estimated timing error

completeness, Fig. 2.21 shows the LMS convergence behavior, along with the time and gain offsets estimation for the sine waves spectrum case. The μ value is the same used in the simulation of Fig. 2.18.

2.5 Recap

In this chapter, the issue of selecting the appropriate sampling rate for a software defined radio receiver has been addressed. An alias-free condition that, if satisfied, guarantees no loss of receiver performance at low-rates in comparison to the analog, or the high rate discrete counterparts has been derived. The mathematical formulation has been derived for a general expression of the modulated signal and a computable formula has been provided for accurately checking the alias-free conditions in the multi-band signal case.

In the second section of this chapter we have studied the timing jitter problem. It can strongly affect the information recovery in a fully digital receiver. A model for taking into account the timing jitter effects has also been derived. Simulation results have been shown for a digital OFDM system. A compact matrix formulation has been used for describing the complete system.

In the last section of the chapter a solution for adaptively canceling time and gain mismatches in interleaved ADCs, that run on communication signals, has been proposed. Time interleaving ADCs is a good solution for overcoming the limits derived from the current ADC hardware technology. By using more ADCs, working in parallel, the sampling rate can be arbitrarily increased. Simulation results have been given in order to verify theoretical results.

References

1. E. Buracchini, CSELT, "The Software Radio Concept," *IEEE Communications Magazine*, vol. 38, no. 9, 2000.
2. J. Mitola, "The Software Radio Architecture," *IEEE Communications Magazine*, vol. 33, no. 5, 1995.
3. f. harris and W. Lowdermilk, "Software Defined Radio: a Tutorial," *IEEE Instrumentation and Measurement Magazine*, vol. 13, no.1, 2010.
4. B. Le, T. W. Rondeau, J. H. Reed, C. W. Bostian, "Analog-to-Digital Converters: a Review of the Past, Present and Future," *IEEE Signal Processing Magazine*, vol. 22, no. 6, 2005.
5. Rodney G. Vaughan, Neil L. Scott, and D. Rod White, "The Theory of Bandpass Sampling," *IEEE Transaction on Signal Processing*, vol. 39, no. 9, 1991.
6. F. Palmieri, G. Romano and E. Venosa, "QAM Receiver with Band-Pass Sampling and Blind Synchronization," *Proc. of 2009 IEEE Aerospace Conference*, March 7-14, 2009, Big Sky, Montana.
7. P. L. Dragotti, M. Vetterli, and T. Blu, "Sampling Moments and Reconstructing Signals of Finite Rate of Innovation: Shannon Meets Strang-Fix," *IEEE Trans. on Signal Processing*, vol. 55, no. 5, 2007.
8. H. Landau, "Sampling, Data Transmission and the Nyquist Rate," *Proc. of IEEE*, vol. 65, no. 10, 1967.
9. M. Unser, "Sampling-50 Years After Shannon," *Proc. of IEEE*, vol. 88, no. 4, 2000.
10. Y. C. Eldar, and T. Michaeli, "Beyond Bandlimited Sampling: A Review of Nonlinearities, Smoothness and Sparsity," *IEEE Signal Proc. Magazine*, vol. 26, no. 3, 2009.
11. D. A. Linden, "A Discussion of Sampling Theorem," *Proceedings of IRE*, vol. 47, no. 7, 1959.
12. A. J. Jerry, "The Shannon Sampling Theorem-Its Various Extensions and Applications: A Tutorial Review," *Proc. of IEEE*, vol. 65, no. 11, 1977.
13. Y. P. Lin and P. P. Vaidyanathan, "Periodically Nonuniform Sampling of Bandpass Signals," *IEEE Trans. Circuits Syst. II*, vol. 45, no. 3, 1998.
14. C. Herley and P. W. Wong, "Minimum Rate Sampling and Reconstruction of Signals with Arbitrary Frequency Support," *IEEE Trans. Inform. Theory*, vol. 45, no. 5, 1999.
15. R. Venkataramani and Y. Bresler, "Perfect Reconstruction Formulas and Bounds on Aliasing Error in Sub-Nyquist Nonuniform Sampling of Multiband Signals," *IEEE Trans. Inform. Theory*, vol. 46, no. 6, 2000.
16. M. Mishali, and Y. C. Eldar, "Reduce and Boost: Recovering Arbitrary Sets of Jointly Sparse Vectors," *IEEE Trans. on Signal Processing*, vol. 56, no. 10, 2008.
17. Y. M. Lu, and M. N. Do, "A Theory for Sampling Signals From a Union of Subspaces," *IEEE Trans. on Signal Processing*, vol. 56, no. 6, 2008.

18. Y. M. Lu, M. N. Do, and R. S. Laugesen, "A Computable Fourier Condition Generating Alias-Free Sampling Lattices," *IEEE Trans. Signal Processing*, vol. 57, no. 5, 2009.
19. M. Mishali, Y. C. Eldar, "Blind Multiband Signal Reconstruction: Compressed Sensing for Analog Signals," *IEEE Trans. on Signal Processing*, vol. 57, no. 3, 2009.
20. D. M. Akos, M. Stockmaster, J. B. Y. Tsui, and J. Caschera, "Direct Bandpass Sampling of Multiple Distinct RF signals," *IEEE Trans. on Communications*, vol. 47, no. 7, 1999.
21. S. Bose, V. Khaitan, and A. Chaturvedi, "A Low-Cost Algorithm to Find the Minimum Sampling Frequency for Multiple Bandpass Signals," *IEEE Signal Processing Letters*, vol. 15, 2008.
22. J. Bae and J. Park, "An Efficient Algorithm for Bandpass Sampling of Multiple RF Signals," *IEEE Signal Processing Letters*, vol. 13, no. 4, 2006.
23. F. Palmieri, E. Venosa, A. Petropulu, G. Romano, P. Salvo Rossi, "Sparse Sampling for Software Radio Receivers", *Proc. of SPAWC 2010 - 11th IEEE International Workshop on Signal Processing Advances in Wireless Communications*, Marrakech, Morocco, 20-23 June, 2010.
24. K. N. Manoj, G. Thiagarajan, "The Effect of Sampling Jitter in OFDM Systems," *IEEE International Conference on Communications 2003*, Anchorage, Alaska, 11-15 May, 2003.
25. V. Vasudevan, "Simulation of the Effects of Timing Jitter in Track-and-Hold and Sample-and-Hold Circuits," *Proc. of 42-nd Design Automation Conference,*, Anaheim, California, 13-17 June, 2005.
26. A. G. Vostretsov, "The Effect of Timing Jitter in Sampling Inside Discrete Systems with High-stable Synchronizing Clock: Models and Analysis," *Proc. of the 4th Korea-Russia International Symposium on Science and Technology, 2000*, Ulsan, Korea, June 27-July 1, 2000.
27. L. Angrisani, M. D'Apuzzo, M. D'Arco, "Modelling Timing Jitter Effects in Digital-to-Analog Converters," *Intelligent Signal Processing, 2005 IEEE International Workshop*, Budapest, Hungary, September 1-3, 2005.
28. A. V. Balakrishnan, "On the Problem of Timing Jitter in Sampling," *IEEE Trans. Inf. Theory*, vol. 8, no. 3, 1962.
29. F. Palmieri, G. Romano and E. Venosa, "A Jitter Model for OFDM," *Proc. CISS 2008 - Conference on Information Sciences and Systems*, Princeton, NJ March 19-21, 2008.
30. fred harris, *Multirate Signal Processing*, Prentice Hall, 2004.
31. S. Saleem and C. Vogel, "LMS-based Identification and Compensation of Timing Mismatches in a Two-Channel Time-Interleaved Analog-to-Digital Converter, *Proc. IEEE Norchip Conf.*, Aalborg, Denmark, November 19-20, 2007.
32. C. Vogel, S. Saleem, and S. Mendel, "Adaptive Blind Compensation of Gain and Timing Mismatches in M-Channel Time-Interleaved ADCs, *Proc. 14th IEEE ICECS*, Marrakech, Morocco, 11-14 December, 2007.
33. S. Huang, B. C. Levy, "Adaptive Blind Calibration of Timing Offset and Gain Mismatch for Two-Channel Time-Interleaved ADCs, *EEE Trans. on Circuits and Systems-I*: Regular Papers: vol. 53, no. 6, 2006.
34. P. Satarzadeh, B. C. Levy, and P. J. Hurst, "Adaptive Semiblind Calibration of Bandwidth Mismatch for Two-Channel Time-Interleaved ADCs, *IEEE Trans. on Circuits and Systems-I*: Regular Papers: vol. 56, no. 9, 2009.

Chapter 3
Radio Design

3.1 Introduction

The issue of optimizing the use of the radio spectrum is becoming more and more pressing because of the growing deployment of new wireless devices and applications. The current inefficient usage of the limited spectrum resources requires changes in the spectrum allocation policy and urges the development of innovative communication technologies that use it in a more intelligent and flexible way.

A software radio becomes cognitive when it acquires the capability to optimally adapt its operating parameters according to the interactions with the surrounding radio environment. This definition implies, on the transmitter side, the capability of the radio to detect the available spectral holes in the spanned frequency range and to dynamically use them for sending signals, having different bandwidths, at randomly located center frequencies.

On the other side of the communication chain, a cognitive receiver has to be able to simultaneously detect multiple signals and down convert them independently of their bandwidths or center frequencies.

Many contributions have appeared in the literature on both particular and generic issues related to cognitive and software defined radio and this proves the interest of the research community on the proposed topics. However a complete radio design has never appeared in the literature until now.

In this chapter, we propose a novel synthesis-analysis chain for cognitive radio that is able to transmit and receive, multiple signals with variable bandwidths, over randomly located center frequencies, according to the temporary radio spectrum availability. Its total workload is kept small by using the polyphase channelizer as the main component in both the receiver and the transmitter. That makes the design immediately suitable for implementation.

The core of proposed transmitter is a synthesis channelizer. It is a variant of the standard M-path polyphase up converter channelizer that is able to perform 2-to-M up sampling while shifting, by aliasing, all the base-band channels to desired center frequencies. Input signals with bandwidths wider than the

E. Venosa et al., *Software Radio: Sampling Rate Selection, Design and Synchronization*, Analog Circuits and Signal Processing, DOI 10.1007/978-1-4614-0113-1_3,

synthesis channelizer bandwidth are pre-processed through small down converter channelizers that disassemble their bandwidths into reduced bandwidth sub-channels. The proposed transmitter avoids the sampled data section replication for multiple simultaneous transmissions and it also allows partitioning and reassembling, when necessary, of the signal spectra before transmitting them.

On the other side of the communication chain, a polyphase analysis channelizer is the key element of the proposed receiver. This engine is able to perform M-to-2 down sampling while simultaneously demodulating, by aliasing, all the received signal spectra having arbitrary bandwidths residing on arbitrary center frequencies. Post-processing up converter channelizers are used for reassembling, from the analysis channelizer base-line channels, signal bandwidths wider than the analysis channelizer channel bandwidth.

3.2   Basics of Multirate Signal Processing

Sample rate change is the central concept in multirate signal processing.

A resampled time series contains samples of the original input time series separated by a set of zero valued samples. The zero valued time samples can be the result of setting a subset of input sample values to zero or the result of inserting zeros between existing input sample values. These two processes are called *down sampling* and *up sampling* respectively.

3.2.1   Multirate Filters

Multirate filters are digital filters that contain a mechanism to increase or decrease the sample rate while processing input sampled signals.

The simplest such filter performs integer up sampling of 1-to-P or integer down sampling of Q-to-1. Of course a multirate filter can employ both up sampling and down sampling in the same process to affect a rational ratio sample rate change of P-to-Q. The integers P and Q may be selected to be the same so that there is no sample rate change between input and output but rather an arbitrary time shift, or phase offset, between input and output sample positions of the complex envelope. The sample rate change can occur at a single location in the processing chain or can be distributed over several sections.

A number of symbols have been used to represent the down sampling and the up sampling elements in a block diagram. Three of the most common are shown in Fig. 3.1.

Note that because the up sampler and down sampler are dual the systems that employ them for sample rate changes will also be seen to be dual.

Conceptually, the process of down sampling can be visualized as a two-step progression as indicated in Fig. 3.2. There are three distinct signals associated with

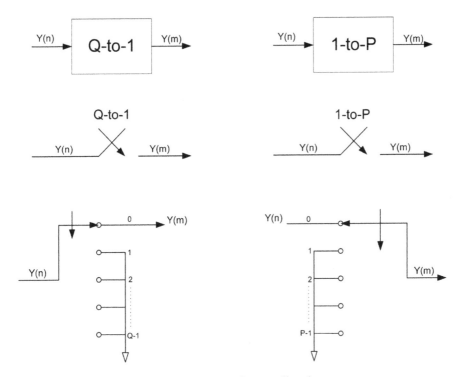

Fig. 3.1 Symbols representing down sampling and up sampling elements

Fig. 3.2 Down sampling process filtering and sample rate reduction

this procedure. The process starts with an input series $x(n)$ that is processed by a filter $h(n)$ to obtain the output sequence $y(n)$ with reduced bandwidth. The sample rate of the output sequence is then reduced to a rate commensurate with the reduced signal bandwidth. In reality the processes of bandwidth reduction and sample rate reduction are merged in a single process called multirate filtering. The bandwidth reduction performed by the digital filter can be a low pass process or a band-pass process.

In a dual way, the process of up sampling can also be visualized as a two-step process as indicated in Fig. 3.3. Here too there are three distinct time series. The process starts by increasing the sample rate of an input series $x(n)$ by resampling it 1-to-P. The zero-packed time series with P-fold replication of the input spectrum is processed by a filter $h(n)$ to reject the spectral replicas and output the sequence $y(m)$ with the same spectrum as the input sequence but sampled at the P-times

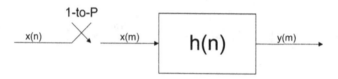

Fig. 3.3 Up sampling process sample rate increasing and filtering

Fig. 3.4 Standard single channel down converter

higher sample rate. In reality also the processes of sample rate increasing and selected bandwidth rejection are merged in a single process again called a multirate filtering.

3.2.2   From Single Channel Down Converter to Channelizer

In this section we address the problem of down converting a single channel of a multi channel frequency division multiplexed (FDM) signal in input to our system. The channels have equal bandwidths and their center frequencies are equally spaced. The input signal has been band limited by analog filters and has been sampled at a sufficiently high sample rate to satisfy the Nyquist criterion for the full FDM bandwidth.

The standard process to down convert a selected channel is shown in Fig. 3.4; the structure shown in Fig. 3.4 performs the standard operations of down conversion of a selected frequency band with a complex heterodyne, low pass filtering, to reduce the bandwidth to the channel, and down sampling to a reduced rate commensurate with the reduced bandwidth. The structure of this processor is seen to be a digital signal processor implementation of a prototype analog I-Q down converter.

The expression for $y(n, k)$, the time series output from the down converted kth channel, prior to resampling, is a simple convolution as shown in (3.1).

$$y(n, k) = [x(n)e^{-j\theta_k n}] * h(n) = \sum_{r=0}^{N-1} x(n-r)e^{-j\theta_k(n-r)}h(r). \qquad (3.1)$$

The output data from the complex mixer is complex, hence is represented by two time series, $I(n)$ and $Q(n)$.

The filter with real impulse response $h(n)$ is implemented as two identical filters, each processing one of the quadrature time series. The convolution process is often

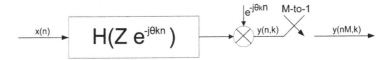

Fig. 3.5 Band-pass version of down converter

performed by a simple digital filter that performs the multiply and adds operations between data samples and filter coefficients extracted from two sets of addressed memory registers. In this form of the filter, one register set contains the data samples while the other contains the coefficients that define the filter impulse response.

We can rearrange the summation of (3.1) to obtain a related summation reflecting the equivalency theorem. This theorem states that

The operations of down conversion followed by a low pass filter are equivalent to the operations of a band-pass filter followed by a down conversion.

Figure 3.5 shown the block diagram of Fig. 3.4 after the application of the equivalence theorem, while the rearranged version of (3.1) is shown in (3.2).

$$y(n, k) = \sum_{r=0}^{N-1} x(n - r)e^{-j\theta_k(n-r)}h(r) =$$

$$= \sum_{r=0}^{N-1} x(n - r)e^{-j\theta_k n}h(r)e^{j\theta_k r} =$$

$$= e^{-j\theta_k n} \sum_{r=0}^{N-1} x(n - r)h(r)e^{j\theta_k r} \tag{3.2}$$

Note that the up converted filter, $h(n)e^{-j\theta_k n}$, is complex and as such its spectrum resides only on the positive frequency axis without a negative frequency image. This is not a common structure for an analog prototype because of the difficulty of forming a pair of analog quadrature filters exhibiting a 90° phase difference across the filter bandwidth.

Applying the transformation suggested by the equivalency theorem to an analog prototype system does not make sense since it doubles the required hardware. We would have to replace a complex scalar heterodyne (two mixers) and a pair of low-pass filters with a pair of band-pass filters, containing twice the number of reactive components, and a full complex heterodyne (four mixers). If it makes no sense to use this relationship in the analog domain, why does it make sense in the digital world? The answer is found in the fact that we define a digital filter as a set of weights stored in coefficient memory. Thus, in the digital world, we incur no cost in replacing the pair of low-pass filters $h(n)$ required in the first option with the pair of band-pass filters $h(n)cos(n\theta_k)$ and $h(n)sin(n\theta_k)$ required for the second one. We accomplish this task by a simple download to the coefficient memory. We still have to justify the full complex heterodyne required for the down conversion at the filter output rather than at the filter input.

Examining Fig. 3.5, we note that following the output down conversion, we perform a sample rate reduction in which we retain one sample in every M-samples. Recognizing that there is no need to down convert the samples we discard in the

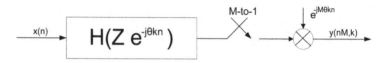

Fig. 3.6 Down sampled band-pass down converter

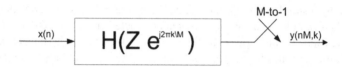

Fig. 3.7 Band-pass down converter aliased to base-band by down sampler

down sample operation, we choose to down sample only the retained samples. This is shown in Fig. 3.6.

Here we note that when we bring the heterodyne to the low data rate side of the resampler, we are also down sampling the time series of the complex sinusoid. The rotation rate of the sampled complex sinusoid is θ_k and $M\theta_k$ radians per sample at the input and output respectively of the M-to-1 resampler. This change in observed rotation rate is due to aliasing. When aliased, a sinusoid at one frequency or phase slope appears at another phase slope due to the resampling. At this point we invoke a constraint on the sampled data center frequency of the down converted channel. We choose center frequencies θ_k, which will alias to DC (zero frequency) as a result of the down sampling to $M\theta_k$. This condition is assured if $M\theta_k$ is congruent to 2π, which occurs when $M\theta_k = k2\pi$, or more specifically, when $\theta_k = k2\pi/M$. The modification to Fig. 3.6 that reflects this provision is seen in Fig. 3.7.

The constraint, that the center frequencies are limited to integer multiples of the output sample rate, assures aliasing to base-band by the sample rate change. When a channel aliases to base-band by the resampling operation the related resampled heterodyne defaults to a unity-valued scalar, which consequently is removed from the signal processing path. If the center frequency of the aliased signal is offset by $\Delta\theta\ rad/smpl$ from a multiple of the output sample rate, the aliased signal will reside at an offset of $\Delta\theta\ rad/sample$ from zero frequency to base-band. A complex heterodyne or base-band converter will shift the signal by the residual $\Delta\theta$ offset. This base-band mixer operates at the output sample rate rather than at the input sample rate for a conventional down converter. We can consider this required final mixing operation a post conversion task and allocate it to the next processing block.

From Fig. 3.7, we note that the current configuration of the single channel down converter involves a band-pass filtering operation followed by a down sampling of the filtered data to alias the output spectrum to base-band. Following the idea that led us to down convert only those samples retained by the down sampler, we similarly conclude that there is no need to compute the output samples from the band-pass filter that will be discarded by the down sampler.

Conceptually we accomplish this in the following manner: after computing an output sample from the non-recursive filter we shift the data in the filter register M positions and wait until M new inputs are delivered to the filter before computing the next output sample. In a sense, by moving input data through the filter in stride of length M we are performing the resampling of the filter at the input port rather than at the output port. Performing the resampling at the filter input essentially reorders the standard operations of the filter followed by a down sample with the operations of down sample followed by the filter. The formal description of the process that accomplishes this interchange is known as the noble identity. This theorem states that

The output from a filter $H(Z^M)$ followed by an M-to-1 down sampler is identical to an M-to-1 down sampler followed by the filter $H(Z)$.

The Z^M in the filter impulse response tells us that the coefficients in the filter are separated by M samples rather than the more conventional one sample delay between coefficients in the filter $H(Z)$. We must take care to properly interpret the operation of the M-to-1 down sampler. The interpretation is that the M-to-1 down sampled time series from a filter processing every Mth input sample presents the same output by first down sampling the input by M-to-1 to discard the samples not used by the filter when computing the retained output samples and then operating the filter on only the retained input samples.

We might ask under what condition does a filter manage to operate on every Mth input sample. We answer by rearranging the description of the filter to establish this condition so that we can invoke the noble identity. This rearrangement starts with an initial partition of the filter into M-parallel filter paths.

The Z-transform description of this partition is presented in (3.3) through (3.6).

$$H(Z) = \sum_{n=0}^{N-1} h(n)Z^{-n} =$$

$$h(0) + h(1)Z^{-1} + h(2)Z^{-2} + h(3)Z^{-3} + \cdots + h(N-1)Z^{N-1}. \quad (3.3)$$

We first examine the base-band version of the noble identity and then trivially extend it to the pass band version.

Anticipating the M-to-1 resampling, we partition the sum shown in (3.3) to a sum of sums as shown in (3.4). In this mapping we load an array by columns but process the array by rows. In our example, the partition forms columns of length M containing M successive terms in the original array, and continues to form adjacent M-length columns until we account for all the elements of the original one-dimensional array.

$$H(Z) = h(0) + h(M+0)Z^{-M} + h(2M+0)Z^{-2M} + \cdots$$

$$h(1)Z^{-1} + h(M+1)Z^{-(M+1)} + h(2M+1)Z^{-(2M+1)} + \cdots$$

$$h(2)Z^{-2} + h(M+2)Z^{-(M+2)} + h(2M+2)Z^{-(2M+2)} + \cdots$$

$$h(3)Z^{-3} + h(M+3)Z^{-(M+3)} + h(2M+3)Z^{-(2M+3)} + \cdots$$

$$\cdots \qquad \cdots$$

$$h(M-1)Z^{M-1} + h(2M-1)Z^{-(2M-1)} + h(3M-1)Z^{-(3M-1)} + \cdots \quad (3.4)$$

We note that the first row of the two-dimensional array is a polynomial in Z^M, which we denote $H_0(Z^M)$, a notation to be interpreted as an addressing scheme to start at index 0 and increments in stride of length M. The second row of the same array, while not a polynomial in Z^M, is made into one by factoring the common Z^{-1} term and then identifying this row as $Z^{-1}H_1(Z^M)$. It is easy to see that each row of (3.4) can be described as $Z^{-r}H_r(Z^M)$ so that (3.4) can be rewritten in a compact form as shown in Equation (3.5).

$$H(Z) = H_0(Z^M) + Z^{-1}H_1(Z^M) + Z^{-2}H_2(Z^M) +$$

$$Z^{-3}H_3(Z^M) + \cdots + Z^{-(M-1)}H(M-1)(Z^M). \quad (3.5)$$

We rewrite (3.5) in the traditional summation form as shown in (3.6), which describes the original polynomial as a sum of delayed polynomials in Z^M.

$$H(Z) = \sum_{r=0}^{M-1} Z^{-r}H_r(Z^M) = \sum_{r=0}^{M-1} Z^{-r} \sum_{n=0}^{N/M-1} h(r+nM)Z^{-nM}. \quad (3.6)$$

The block diagram reflecting this M-path partition of a resampled digital filter is shown in Fig. 3.8.

The output formed from the M separate filter stages representing the M separate paths is the same as that obtained from the non partitioned filter. We have not yet performed the interchange of filter and resampling. We first pull the resampler through the output summation element and down sample the separate outputs by performing the output sum only for the retained output sample points. With the resamplers at the output of each filter, which operates on every Mth input sample, we are prepared to invoke the noble identity and pull the resampler to the input side of each filter stage. This is shown in Fig. 3.9.

The input resamplers operate synchronously, all closing at the same clock cycle. When the switches close, the signal delivered to the filter on the top path is the current input sample. The signal delivered to the filter one path down is the content

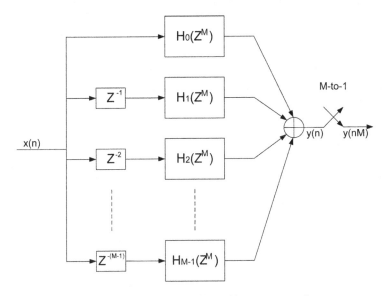

Fig. 3.8 M-path partition of prototype low-pass filter with output resampler

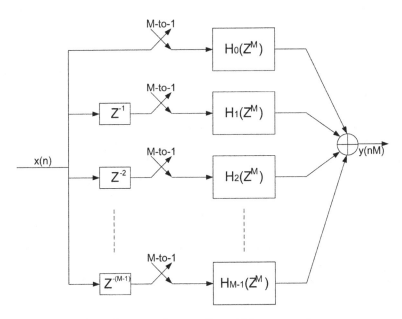

Fig. 3.9 M-path partition of prototype low-pass filter with input resamplers

of the one sample delay line, which of course is the previous input sample. Similarly, as we traverse the successive paths of the M-path partition, we find upon switch closure that the kth path receives a data sample delivered k-samples ago.

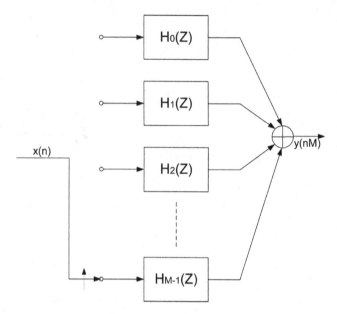

Fig. 3.10 M-path partition of prototype low-pass filter with input path delays and M-to-1 resamplers replaced by input commutator

We conclude that the interaction of the delay lines in each path with the set of synchronous switches can be likened to an input commutator that delivers successive samples to successive legs of the M-path filter. This interpretation is shown in Fig. 3.10.

We now complete the transformation that changes a standard mixer down converter to a resampling M-path down converter: we apply the frequency translation property of the Z-transform illustrated in (3.7)–(3.9).

If

$$H(Z) = h(0) + h(1)Z^{-1} + h(2)Z^{-2} + h(3)Z^{-3} + \cdots + h(N-1)Z^{-(N-1)} \quad (3.7)$$

and

$$G(Z) =$$

$$= h(0) + h(1)e^{j\theta}Z^{-1} + h(2)e^{j2\theta}Z^{-2} + \cdots + h(N-1)e^{j(N-1)\theta}Z^{-(N-1)} =$$

$$= h(0) + h(1)e[j\theta Z]^{-1} + h(2)[e^{j\theta}Z]^{-2} + \cdots + h(N-1)[e^{j\theta}Z]^{-(N-1)} \quad (3.8)$$

then

$$G(Z) = H(Z)|_{Z \Rightarrow e^{-j\theta}Z} = H(e^{-j\theta}Z). \quad (3.9)$$

Equations (3.7)–(3.9) proof that: if $h(n)$, the impulse response of a base-band filter, has a Z-transform $H(Z)$, then the sequence $h(n)e^{j\theta n}$, the impulse response of a

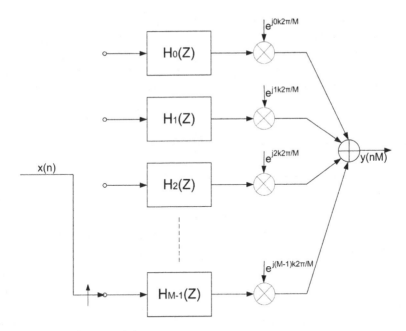

Fig. 3.11 Resampling M-path down converter

band-pass filter, has a Z-transform $H(Ze^{-j\theta_n})$. We can convert a low-pass filter to a band-pass filter by associating the complex heterodyne terms of the modulation process of the filter weights with the delay elements storing the filter weights.

We now apply this relationship to (3.2) by replacing each Z^{-1} with $Z^{-1}e^{j\theta_n}$, with the phase term θ satisfying the congruency constraint that $\theta = k(2\pi/M)$. Thus Z^{-1} is replaced with $Z^{-1}e^{jk(2\pi/M)}$, and Z^{-M} is replaced with $Z^{-M}e^{jkM(2\pi/M)}$.

By design, the k-th multiple of $2\pi/M$ is a multiple of 2π for which the complex phase rotator term defaults to unity, or in our interpretation, aliases to base-band. The default to unity of the complex phase rotator occurs in each path of the M-path filter shown in Fig. 3.10.

The non default complex phase angles are attached to the delay elements on each of the M paths. For these delays, the terms Z^{-r} are replaced by the terms $Z^{-r}e^{jkr(2\pi/M)}$. The complex scalar $e^{jkr(2\pi/M)}$ attached to each path of the M-path filter can be placed anywhere along the path, and in anticipation of the next step, we choose to place the complex scalar after the down sampled path filter segments $H_r(Z)$. This is shown in Fig. 3.11.

When the polyphase filter is used to down convert and down sample a single channel, the phase rotators are implemented as external complex products following each path filter.

When a small number of channels (less than 10) are being down converted and down sampled appropriate parallel sets of phase rotators can be applied to the filter stage outputs and summed to form each channel output.

When the number of channels becomes on the order of $log2(N)$, since the phase
rotators following the polyphase filter stages are the same as the phase rotators of a
inverse discrete Fourier transform, we can use the IDFT to simultaneously apply the
phase shifters for all of the channels we wish to extract from the aliased signal set.
Obviously for computational efficiency, the IFFT algorithm implements the IDFT.

3.2.3   M-to-2 Down Converter Channelizer

We can describe the three basic operations performed by a standard polyphase
M-to-1 down converter as:

- Sample rate change, due to the input commutator
- Bandwidth reduction, due to the M-path partitioned filter weights
- Nyquist zone selection, due to the IFFT block

These three operations are completely independent of each other and they can
be modified to achieve different goals for both the receiver and transmitter. The
standard down converter channelizer is critically sampled with channel spacing,
channel bandwidth, and output sample rate equal to $\frac{f_s}{M}$. This causes the transition
band edges of the channelizer filters to alias onto itself which would prevent the use
of its output signal in following processing steps.

In order to solve this problem, we can modify the basic channelizer design
doubling its output sample rate without changing the channel bandwidth or channel
spacing. In doing so, the sample rate will be twice the bandwidth of the prototype
filters as well as twice the channel spacing. The reconfigured version of the standard
M-path polyphase down converter channelizer is shown in Fig. 3.12. It is capable
of performing the sample rate change from the input rate f_s to the output rate $\frac{2f_s}{M}$.
Its detailed derivation can be found in the references at the end of this chapter. In
this section we briefly explain the way in which it works.

The input data buffer performs the correct data loading of the $\frac{M}{2}$ input samples
into the M-path filter while the circular output buffer performs the time alignment
of the shifting time origin of the input samples in the M-path filter with the non
shifting time origin of the phase rotator outputs of the IFFT. The $\frac{M}{2}$ time sample
shift of the input time series redefines the time origin and causes sinusoids with an
odd number of cycles in the length M array to alternate sign on successive shifts.
The alternating sign is the reason that the odd indexed frequency bins alias to the
half sample rate while the even indexed frequency bins alias to DC. Rather than
reverse phase of alternate output samples from the odd indexed bins we perform an
$\frac{M}{2}$ point circular shift of alternate M-length vectors before presenting the vector to
the IFFT. The circular shift applies the correct phase alignment to all frequencies
simultaneously. Of course, there is not a circular shift of the buffer but rather an
alternate data load into the IFFT input buffer.

Note that if we are thinking to use the M-to-2 down converter channelizer
in a software radio transceiver we have to pay particular attention in designing

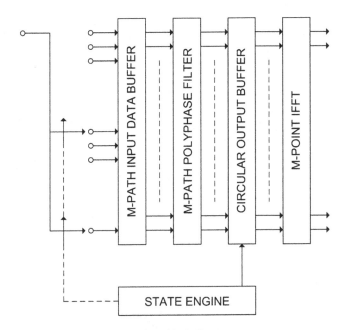

Fig. 3.12 M-to-2 down converter channelizer block diagram

its low-pass prototype filter. In a SDR transceiver, the signal spectra will be randomly located in the frequency domain and their bandwidths will easily span and occupy more than one base-line channel. We need to collect and process all the spanned channels corresponding to a single input bandwidth; we need to fragment and defragment them, even more than once during the processing chain without processing artifacts at the output of the synthesizer block. That is more than a good reason for using SQRT-Nyquist filter as low-pass prototype in the modified version of the standard M-to-1 down converter channelizers. By doing so we place M of them across the whole spanned spectrum with each filter centered on $\frac{kf_s}{M}$. All adjacent filters exhibit $-3\,$dB overlap at their band edges. The channelizer working under this configuration is able to collect all the signals' energy across its full operating spectrum range even if signals occupy more than one adjacent channel and resides in the channel's overlapping transition bandwidths.

Note that in a SDR the SQRT-Nyquist prototype low-pass filter has to be designed with its two sided 3 dB bandwidth equal to $\frac{1}{M}$-th of the channelizer input sampling frequency. This is equivalent to the filter impulse response having approximately M samples between its peak and first zero crossing and having approximately M samples between its zero crossings. The integer M is also the size of the IFFT as well as the number of channels in the M-to-2 channelizer. The prototype filter must also exhibit reasonable transition bandwidth and sufficient out of band attenuation or stop-band level. We designed our system for a dynamic range of 80 dB, the dynamic range of a 16-bit processor.

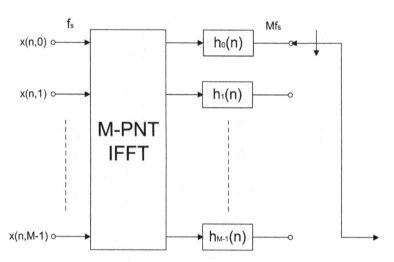

Fig. 3.13 Standard M-path polyphase modulator: M-point IFFT, M-path polyphase filter and M-port commutator

3.2.4   2-to-M Up Converter Channelizer

The standard 1-to-M up converter channelizer simultaneously performs three tasks:

- Up conversion to the selected Nyquist zones, due to the IFFT block
- Filtering, due to the M-path partitioned filter weights
- Sample rate change, due to the output commutator

These three operations are completely independent of each other and they can be modified to achieve different goals.

Figure 3.13 shows the complete structure of the standard 1-to-M up converter channelizer formed by an M-point inverse discrete Fourier transform, an M-path partitioned low-pass prototype filter and an M-port commutator.

In this engine, M-point IFFT performs two simultaneous tasks: an initial 1-to-M up sampling that forms an M-length vector for each input sample $x(n, k)$ and, further a complex phase rotation of k cycles in M-samples on the up sampled output vector. The IFFT generates a weighted sum of complex vectors containing integer number of cycles per M-length vector. The partitioned polyphase filter forms a sequence of column coefficient weighted, MATLAB's dot-multiply, versions of these complex spinning vectors. The sum of these columns, formed by the set of inner products in the polyphase partitioned filter, is the shaped version of the up converted M-length vector output from the IFFT. The M-port output commutator finally, takes M consecutive samples from the output ports of the M-path filter to deliver the 1-to-M interpolated, up converted and shaped time series formed by the synthesizer channelizer.

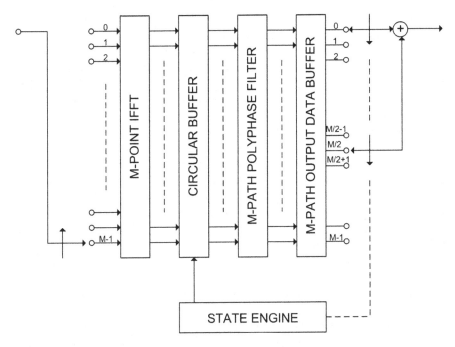

Fig. 3.14 2-to-M up converter channelizer block diagram

Note that, as the standard down converter channelizer, the standard up converter channelizer is critically sampled when channel spacing, channel bandwidth, and input sample rate are all equal to f_s. The input signals to the up converter channelizer are externally shaped. The sample rate of the shaped signal must, of course, satisfy the Nyquist criterion. We often choose to operate shaping filters at 2-samples per symbol for ease of signal generation while satisfying the Nyquist criterion. The minimum channel center positioning of a channelizer should be matched to the two-sided bandwidth of the input signal, not to its symbol rate or twice its symbol rate.

   To easily accommodate the channel spacing requirement, we modify the basic channelizer to require the input sample rate to be twice the two-sided bandwidth of the input signal rate without changing the channel bandwidth or channel spacing. In doing so, the input sample rate will be twice the bandwidth of the prototype filters as well as twice the channel spacing. In this configuration, the channelizer is capable of performing the sample rate change from the input rate $2f_s$ to the output rate Mf_s offering the advantage of permitting the input signal to be oversampled by 2 and avoiding the difficulty of having the sample rate precisely matched to the two sided bandwidth of the input signals as well as permitting a shorter length prototype channelizer filter due to an allowable wider transition bandwidth. Here we briefly explain the 2-to-M channelizer block diagram that is shown in the second processing block of Fig. 3.14. More details on it can be found in the references at the end of this chapter.

The M-point IFFT applies the complex phase rotation to the separate base-band input signals as well as performs the initial 1-to-M up sampling of the input samples. The circular buffer following the IFFT performs the correct data offset of the two $\frac{M}{2}$ points at the output of the IFFT for maintaining phase alignment with the $\frac{M}{2}$ channelizer output vector. The complex sinusoid at the output of the IFFT always defines its time origin as the initial sample of the IFFT output vector. The output of the polyphase filter exhibits a time origin that shifts due to the $\frac{M}{2}$ time sample shift embedded in the output commutator.

The $\frac{M}{2}$ time sample shift of the output time series causes sinusoids with an odd number of cycles in the length M array to alternate sign on successive shifts. The alternating sign is the reason that the odd indexed frequency bins up convert to a frequency $k + \frac{N}{2}$ rather than frequency index k. Rather than reverse phase alternate input samples to the odd indexed IFFT bins usually an $\frac{M}{2}$ point circular shift of alternate M-length vectors from the IFFT is performed. The circular shift applies the correct phase alignment to all frequencies simultaneously. Of course, in practice, there is not a circular shift of the buffer but rather an alternate data load from the IFFT output buffer into the polyphase filter.

The same reasoning on the low-pass prototype filter requirements we have done in the previous section, has also to be applied to this modified version of the standard up converter channelizer if we are planning to use it in a software defined radio. The input spectra are placed at randomly located positions in the frequency domain and their bandwidths can easily span and occupy more than one base-line channel and that is the motivation for which we need to use SQRT-Nyquist as prototype low-pass filter. The M-to-2 down converter channelizer working under this configuration is able to collect all the energy of the signals across its full operating spectrum range even if the signals occupy more than one adjacent channel and reside in the channel's overlapping transition bandwidths.

3.2.5 Nyquist and SQRT-Nyquist Filters

We specified in the earlier sections that in order to get the flexibility to demodulate randomly located center frequency signals we have to use SQRT-Nyquist filter as a low-pass prototype in the polyphase channelizer.

In the continuous time domain a Nyquist filter is achieved by multiplying a Nyquist pulse with another pulse shape with finite support showing a symmetric continuous spectrum. A second window, a rectangle, performs the truncation of the Nyquist pulse in the time domain.

Equation (3.10) presents the most common continuous time expression for a Nyquist filter achieved by multiplying the *sinc* function with the *raised cosine* function.

$$h_{NYQ}(t) = f_{SYM} \frac{sin(\pi f_{SYM} t)}{\pi f_{SYM} t} \frac{cos(\pi \alpha f_{SYM} t)}{[1 - (2\alpha f_{SYM} t)^2]} \tag{3.10}$$

Fig. 3.15 Example of channelized spectrum using Nyquist filter

Equation (3.11) shows its frequency response.

$$
H_{NYQ}(w) = \begin{cases} 1 & : \quad for\, \dfrac{|\omega|}{\omega_{SYM}} \le (1-\alpha) \\ 0.5 * \left\{1 + cos\dfrac{\pi}{2\alpha}\left[\dfrac{\omega}{\omega_{SYM}} - (1-\alpha)\right]\right\} & : for(1-\alpha)\dfrac{|\omega|}{\omega_{SYM}} \le (1+\alpha) \\ 0 & : \quad for\, \dfrac{|\omega|}{\omega_{SYM}} \ge (1+\alpha) \end{cases}
$$
(3.11)

Nyquist filter presents the interesting property to show a gain of 0.5 (or $-6\,dB$) at the nominal band edge; this property allows perfect reconstruction also for the signal spectra falling in the transition bands.

Figure 3.15 shows a qualitative example of channelization using Nyquist proto-type filter.

In the channelizers that form both the SDR receiver and transmitter structures presented in the following sections we use SQRT-Nyquist filters ($\sqrt{H_{NYQ}(w)}$) as low-pass prototypes. These filters show a gain of 0.707 (or $-3\,dB$) at the nominal band edge and the composition of two of them (our structures present two tiers channelizer in both the receiver and the transmitter) gives us a Nyquist filter that allows perfect reconstruction of the signals also in the case in which they fall in the filter transition bandwidths.

3.3 Transmitter Design

A digital transmitter has to perform the first tier tasks of filtering, spectral translation and signal conversion which will be matched and reversed by first tier processing performed at the receiver.

Figure 3.16 shows the block diagram of first tier processing of a typical digital transmitter. It accepts binary input sequences and outputs radio frequency (RF) amplitude and phase modulated wave shapes. The digital signal processing (DSP) part of this process starts by accepting b-bit words from a binary source at input symbol rate. These words address a look-up table that outputs gray coded ordered pairs, I-Q constellation points that control the amplitude and phase of the modulated carrier. The I-Q pair is input to digital shaping filters that form 1-to-4 up sampled time series designed to control the wave shape and limit the base-band modulation bandwidth. The time series from the shaping filter are further up sampled by a pair of interpolating filters to obtain a wider spectral interval between sampled data spectral

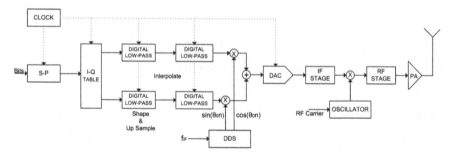

Fig. 3.16 Block diagram of primary signal processing in a typical digital transmitter

replicas. The interpolated data is then heterodyned by a digital up converter to a convenient digital intermediate frequency and then moved from the sampled data domain to the continuous analog domain by a digital to analog converter and analog IF filter. Further analog processing performs spectral transformations required to couple the wave shape to the channel.

Note that to simultaneously up convert multiple signals from base-band using the transmitter structure shown in Fig. 3.16, it is necessary to replicate its sampled data section. Here is the main limitation: the more signals we have to simultaneously up convert, the more sampled data sections we need to implement!

In the following, we propose a novel structure that can be used in a software radio transmitter. By using it we avoid the need to replicate the sampled data section. It is sufficient to use only a single partitioned prototype filter preceded by an IFFT block to simultaneously interpolate and up convert all the transmitter signals. Its structure is based on the polyphase up converter channelizer (synthesis channelizer) and it is able to process all the channels in the output spectral span. Input signals with wider bandwidth than the channelizer bandwidth are easily accommodated. In fact smaller pre processing analysis channelizers will disassemble arbitrary input bandwidths with randomly positioned frequency offsets into reduced bandwidth sub-channels matched to the base-line channelizer bandwidths which are reassembled by the synthesis channelizer.

3.3.1 Proposed Synthesis Channelizer

Figure 3.17 shows the complete block diagram of the proposed synthesis channelizer. It is composed of a 1-to-$\frac{M}{2}$ up converter channelizer preceded by N smaller pre-processing analysis blocks. All the input signals to the 2-to-M synthesis channelizer with bandwidths less than or equal to the channel spacing are to be up sampled and up converted from base-band. For signals with bandwidths less than channel bandwidth we may have to filter and resample with a second synthesis channelizer to obtain the desired input sampling rate of 2-samples per channel

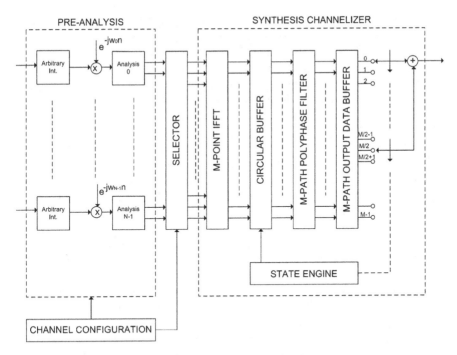

Fig. 3.17 Proposed synthesis channelizer

bandwidth. Moreover, we expect that many of the spectra we presented to the second channelizer in the previous section have bandwidths that are wider than the channel spacing. We must fragment their bandwidth, translate the spectral segment to base-band, and reduce the sample rate to twice the synthesis channelizer channel bandwidth. These are exactly the tasks performed by the small analysis channelizers. They are designed as a P_n-to-2 down converter channelizer where P_n is approximately twice the number of baseline channels spanned by the wideband spectrum. P_n is also the IFFT size and it is chosen to span the bandwidth of the nth input signal spectrum. Their task is to decompose, when necessary, a wider input spectrum into P_n channels matching the bandwidth and sample rate of the baseline synthesis channelizer.

We explain the reason for including the analyzer in our design in this way: all the signals at the input of the 1-to-$\frac{M}{2}$ up converter channelizer with bandwidths smaller than channel spacing are up sampled and their frequency bands are translated from base-band to the selected center frequency by the channelizing process. In order to insert a wider bandwidth signal into the synthesis channelizer we first have to disassemble it into segments with sample rate and bandwidth expected by the synthesis channelizer.

Note that this is the dual processing tasks performed by the previously described up sampling channelizer. This system accepts $\frac{N}{2}$ input samples and outputs time samples from P_n output channels. The N-point input buffer is fed by a dual input commutator with an $\frac{N}{2}$ sample offset. The N-path filter contains polynomials of the form $H_r(Z^2)$ and $Z^{-1}H_{r+N/2}(Z^2)$ in its upper and lower halves respectively. The circular output buffer performs the phase rotation alignment of the IFFT block with the $\frac{N}{2}$ stride shifting time origin of the N-path polyphase filter.

The IFFT size is P_n, this is also the number of the filter output channels. In order to recombine the segmented signal components we have to satisfy the Nyquist criteria for their sum.

Since this is a pre-processing analysis channelizer that feeds the M-path synthesis channelizer we must have its output sample rate to be twice the channel bandwidth symbol rate. We can achieve this by selecting P_n to be approximately twice the number of channels being merged in the synthesizer and setting its input sample rate to be $2f_sP_n$ so that the preprocessor output rate per channel is the required $2f_s$. An actual system may have a few standard size IFFT's to be used for the analysis channelizers and the user may have to choose from the small list of available block sizes. The block sizes must be even to perform the 2-to-N resampling by the technique described here.

The channel selector placed between the analysis bank and the IFFT also connected with the input channel control will provide the correct outputs from the pre processor analysis channelizer to the output synthesizer.

Particular attention has to be paid in designing the low-pass prototype filter used in the modified synthesis channelizer. The input spectra are to be placed at randomly located positions in the frequency domain and their bandwidths can easily span and occupy more than one base-line channel. When an input signal bandwidth extends over multiple baseline channel widths, it is pre-decomposed by an input analysis channelizer that has a number of output channels required to accommodate the entire wider bandwidth. These segmented channels are reassembled without processing artifacts at the output of the synthesizer block by using SQRT-Nyquist filters.

Note that the choice of M, in the synthesis channelizer, determines the spectral resolution of the synthesis channelizer. If multiple narrow band signals fall in the same channel bandwidth, a further filtering process may be required to separate them. By designing the channel spacing to accommodate the most likely expected channel width we minimize the need for the further filtering stage. The optimum choice of M also depends on the channelizer's application.

A channel configuration block, also present in the system, provides the necessary information to a selector block that is connected to the analysis input series. The selector routes all the segments required to assemble the wider bandwidth channel to the synthesizer that performs the reassembly.

Depending on the center frequency at which we desire to shift the signals, a complex frequency rotator block may be placed before the analyzers to allow the completely arbitrary center frequency positioning of the signals which are about to be transmitted.

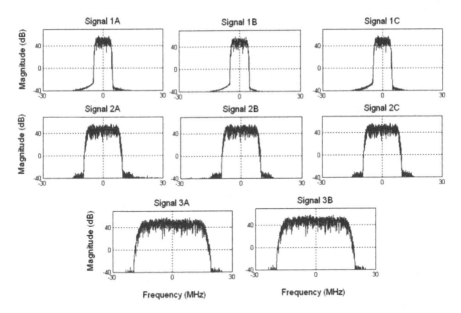

Fig. 3.18 Base-band QPSK signal spectra

3.3.2 Simulation Results

In the simulation results shown in this section we consider a set of distinct input signals delivered to the 2-to-M up converter channelizer. These are QPSK signals with three different bandwidths as shown in Fig. 3.18. The symbol rates chosen for the three signals denoted 1, 2, and 3, are 7.5, 15.0 and 30.0 MHz. The signals are shaped by SQRT-Nyquist filters with 25% excess bandwidth, hence the two-sided bandwidths are $7.5 * 1.25$, $15.0 * 1.25$, and $30.0 * 1.25$ MHz respectively. The IFFT transform size of the base-line 2-to-M up converter channelizer is $M = 48$ with 10 MHz channel spacing for which the required input sample rate per input signal is 20 MHz and for which the output sample rate will be 480 MHz.

For ease of signal generation, all three signals were shaped and up sampled to 60 MHz sample rate with shaping filters designed for 8, 4, and 2 samples per symbol respectively.

Signal 1, the first signal, is down sampled 3-to-1 to obtain the desired 20 MHz sample rate for the up converter channelizer. Signals 2 and 3, the second and third signals are down sampled 6-to-2 in 6 point IFFT analysis channelizers which form 10 MHz channels at 20 MHz sample rate. Signal 2 is spanned by three 10 MHz channels which will feed 3 input ports of the 48 point IFFT and signal 3 is spanned by five 10 MHz channels which will feed five input ports of 48 point IFFT controlled by a channel control block that routes them to the selected synthesis channelizer ports.

Any of the signals presented to the analysis channelizers can be base-band heterodyned or frequency shifted prior to the down sampling channelizer.

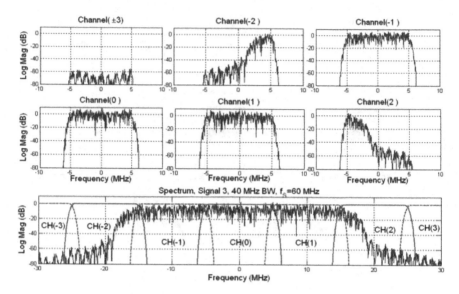

Fig. 3.19 Spectral fragments formed by 6-channel 6-to-2 down sample analysis channelizer processing wide bandwidth signal 3

The output channels that span the offset input bandwidth of the analysis channelizer are the channels passed to the synthesizer and these change due to a frequency offset. The inserted base-band frequency offset will survive the synthesis channelizer.

Figure 3.19 shows all the spectra of the output time series from the 6 channel polyphase 6-to-2 analysis channelizer engine processing signal 3, the wideband signal.

Also seen is the spectrum of signal 3 and the frequency response of the six channels formed by the analyzer.

Note that the spectra in the upper subplots have been filtered by the channelizer 10 MHz SQRT-Nyquist pass-band frequency response, have been translated to baseband and have been sampled at 20 MHz. Five of these segments are presented to the five input ports of the 2-to-48 synthesis channelizer centered on the desired frequency translation index. The synthesizer filters are also SQRT-Nyquist filters which finish shaping the segmented spectral intervals. So they have been SQRT-Nyquist filtered as they are up converted and reassembled by the synthesizer.

The up converter polyphase synthesis channelizer accepts time sequences sampled at 20 MHz with bandwidths less than 10 MHz. We have delivered three signals that satisfy these constraints along with four signals that were conditioned by analysis channelizers that partitioned their bandwidths into segments that also satisfied the input signal constraints.

The spectra of the separate components delivered to the synthesis channelizer are shown in Fig. 3.20. It is easy to recognize in this figure the different spectra

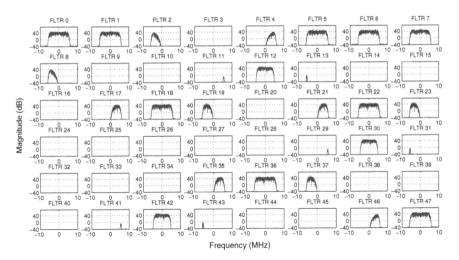

Fig. 3.20 2-to-M up converter channelizer outputs

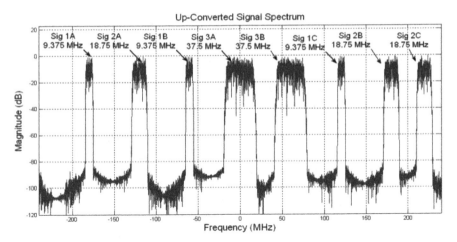

Fig. 3.21 Up converter synthesized spectrum with unequal bandwidth fragmented and defragmented spectral components

forming the received signal. Here we see that filters 4-through-8 are segments of a single frequency band fragmented in Fig. 3.18. At this point, we up sample, frequency shift, and recombine the time series from coupled channel outputs using the synthesis channelizer.

The spectra of the eight channels, including six that have been fragmented in analysis channelizers and then defragmented in the synthesis channelizer are plotted in Fig. 3.21.

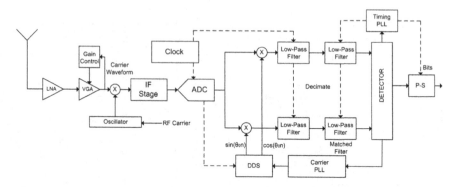

Fig. 3.22 Block diagram of primary signal processing in a typical digital receiver

3.4   Receiver Design

In this section, we present the dual version of the synthesis analysis transmitter chain presented in the previous section.

We recall that a digital receiver has to perform the first tier tasks of filtering, spectral translation and signal conversion to reverse the dual processing tasks performed at the transmitter. The receiver must also perform a number of second tier tasks not present in the transmitter. These tasks are performed to estimate the unknown parameters of the received signal such as amplitude, frequency and timing alignment. Figure 3.22 shows the block diagram of first and second tier processing of a typical digital receiver.

The digital receiver samples the output of the analog intermediate frequency filter and down converts the intermediate frequency centered signal to base-band with a digital down converter. The base-band signal is down sampled by a decimating filter and finally processed in the matched filter to maximize the signal to noise ratio of the samples presented to the detector. The digital signal processing portion of this receiver also includes carrier alignment, timing recovery, channel equalization, automatic gain control, signal to noise estimation, signal detection and interference suppression blocks.

Because the receiver contains analog hardware components the receiver incorporates a number of third tier digital signal processing compensating blocks to suppress the undesired artifacts formed by the analog blocks.

As for the transmitter side, the main limitation of this structure is the replication of the sampled data processing section to simultaneously down convert multiple signals to base-band; the more signals we have to simultaneously down convert, the more sampled data sections we need to implement!

In order to overcome this limitation, in the following we propose a novel analysis channelizer that is able to simultaneously demodulate all the received signals by using only a single partitioned low-pass prototype filter followed by an IFFT block.

3.4.1 Proposed Analysis Channelizer

We recall that, in the standard polyphase down sampling channelizer, the commutator delivers M consecutive samples to the M input ports of the M-path filter performing the signal sample rate reduction which causes M spectral folds in the frequency domain. With an output sample rate of $\frac{f_s}{M}$, all M multiples of the output sample rate alias to base-band (DC). The alias terms in each arm of the M-path filter exhibit unique phase profiles due to their distinct center frequencies and to the time offsets of the different down sampled time series delivered to each commutator port. In particular, each of the aliased terms exhibits a phase shift equal to the product of its center frequency k with its path time delay r. These phase shifts are shown in (3.12).

$$\phi(r, k) = -\omega_k \Delta T_r = -2\pi \frac{f_s}{M} k r T_s = -\frac{2\pi}{M} rk \tag{3.12}$$

where f_s is the sample rate at the input of the polyphase down converter.

The partitioned M-path filter aligns the time origin of the sampled data sequences delivered by the input commutator to a single common output time origin. This task is accomplished by the all-pass characteristics of the M-path partitioned filter that applies the required differential time delay to the individual input time series. Finally the IFFT block performs the equivalent of a beam-forming operation; the coherent summation of the time aligned signals at each output port with selected phase profile. The phase coherent summation of the outputs of the M-path filters separates the various aliases residing in each path by constructively summing the selected aliased frequency components located in each path, while simultaneously destructively canceling the remaining aliased spectral components. The IFFT block extracts, in each arm, from the myriad of aliased signals only the alias with the particular matching phase profile.

Our first change to the basic channelizer is to double the output sample rate without changing the channel bandwidth or channel spacing. In doing so the sample rate will be twice the bandwidth of the prototype filters as well as twice the channel spacing.

Figure 3.23 shows the complete block diagram of the proposed analysis channelizer.

It is composed of an $\frac{M}{2}$-to-1 down sampler channelizer using SQRT-Nyquist prototype filter, followed by N synthesizer blocks. At the output of the first M-to-2 channelizer all the signal bandwidths less than or equal to the channel spacing have been aliased to base band.

For signals with bandwidths less than the analysis channelizer channel bandwidth we have to filter and resample with arbitrary interpolators to obtain desired output sampling rate of 2-samples per symbol bandwidth. However, we expect that many of the spectra we presented to the first channelizer in the previous section have bandwidths which are wider than the channel spacing. The channelizer partitioned their bandwidths into several fragments and aliased every segment to base-band.

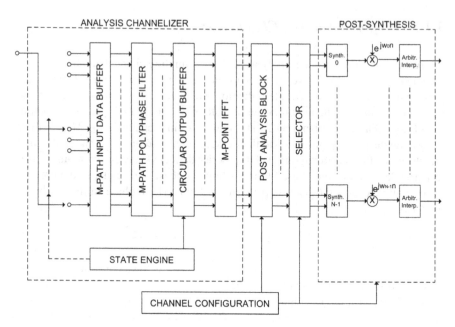

Fig. 3.23 Proposed modified analysis channelizer

In order to recombine the fragments we must, first, up sample each input time series and second, translate them to their proper spectral region. We can then form the sum to obtain the super channel representation of the original signal bandwidth. This is exactly the task performed by the synthesis channelizers. They are designed as a 2-to-P_n up converter channelizers where P_n, the IFFT size, is chosen to span the bandwidth of the nth received signal spectrum. It has to be even and also it has to span enough input channel bands to form an output sample rate of two samples per assembled bandwidth. We note that it is a smaller version of the dual processing tasks performed by the analysis channelizer.

Because we need to collect and process all spanned channels corresponding to a single input bandwidth and assemble them without processing artifacts at the output of the synthesizer block we need to use SQRT-Nyquist filter as low-pass prototype in our channelizers. By doing so we place M of them across the whole spanned spectrum with each filter centered on $\frac{kf_s}{M}$. All adjacent filters exhibit -3 dB overlap at their band edges.

A channel configuration and sub-channel alignment block provides the necessary information to a selector block that is connected to the synthesizer output series. The selector routes all the segments required to assemble the wider bandwidth channel to the synthesizer that performs their reassembly. Depending on the center frequency of the assembled spectrum a further filtering and frequency shifting block may be placed after the synthesizers to shape and properly frequency align and bandlimit the assembled signal.

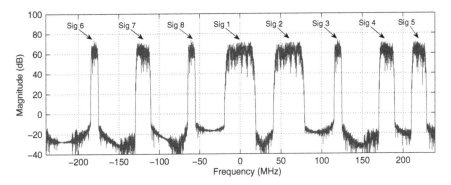

Fig. 3.24 Received signal spectrum

3.4.2 Simulation Results

In the simulation results shown in this section we are considering a composite input signal to the M-to-2 down converter channelizer that contains eight randomly located QPSK signals with three different bandwidths as shown in Fig. 3.24; the signals are sampled at 480 MHz. The bandwidths chosen for these signals are $7.5 * 1.25$ MHz, $15 * 1.25$ MHz, and $30 * 1.25$ MHz. The signals are shaped by SQRT-Nyquist filters with 25% excess bandwidth.

The IFFT transform size of the first M-to-2 down converter channelizer is $M = 48$ with an output sample rate per channel of 20 MHz. We have 48 inputs to the selector block that, controlled by a channel control block, routes them to the selected synthesis channelizers. Because in this example we are using three different bandwidths we could have three different synthesizers. However, because one of the processed bandwidth is smaller than the channel spacing that for this example is 10 MHz, we have no need to deliver these signals to a synthesizer.

The narrowest three bandwidths of the eight shown in Fig. 3.24 are already adequately channelized at the output of the first channelizer; the only task to perform is to resample them in order to achieve the desired 15 MHz output sampling rate. We do have need to process the five remaining bands. The IFFT sizes, P_n with $n = 0, 1$, we selected to process the other two bandwidths are: 4-PNT for the biggest one and 6-PNT for the other one. These sizes are the minimum possible that provide at least two samples per symbol. Note that for the $15 * 1.25$ MHz bandwidth signals, the minimum IFFT size to get two samples per symbol is 3 but, because of the structure of the synthesizers, we can only have an even number of IFFT points. For that reason we choose this number to be 4. It is the smallest possible even index giving us at least two samples per symbol. The sampling rate at the output of this synthesizer is 40 MHz so we simply resample the signals to achieve the desired 30 MHz output rate.

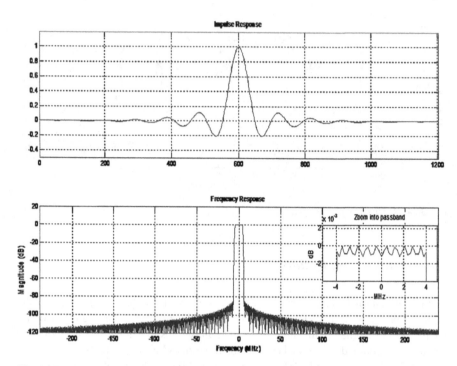

Fig. 3.25 Nyquist low-pass prototype filter

Figure 3.25 shows the impulse response and the magnitude response of the designed prototype low-pass SQRT-Nyquist filter. It is designed to be 48 samples per symbol. Its length is 1,200 taps while its stop band attenuation is -80 dB. Note that, since this filter is M-path partitioned, the length of each filter in the M-path bank is only 25 taps.

Figure 3.26 shows all the spectra of the output time series from the 48 channels of the polyphase M-to-2 analysis channelizer engine. The down converter polyphase channelizer has partitioned the entire spanned frequency range into 48 segments. It is easy to recognize in this figure the different spectra composing the received signal. Here we see that filters 4 through 8 are segments of a single frequency band. At this point, we up sample, frequency shift, and recombine the time series from coupled channel outputs using the synthesis channelizers.

The spectra of the eight channels, including six that have been defragmented are plotted in Fig. 3.27.

We also match-filtered each of the eight channelized and reconstructed time signals and present their constellations in Fig. 3.28. Here we see that all of the QPSK constellations shrink to one point demonstrating the correct functionality of the proposed channelizer.

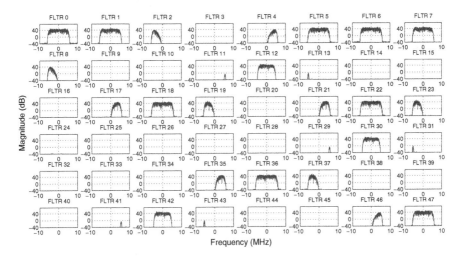

Fig. 3.26 M-to-2 down converter channelizer outputs

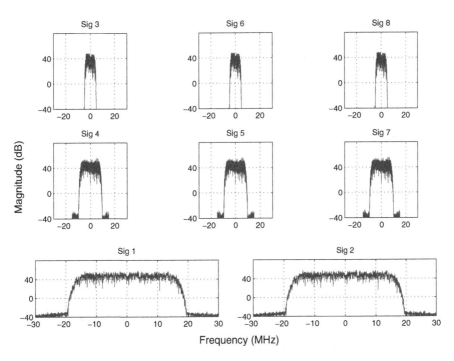

Fig. 3.27 Synthesizer outputs: demodulated and reconstructed spectra

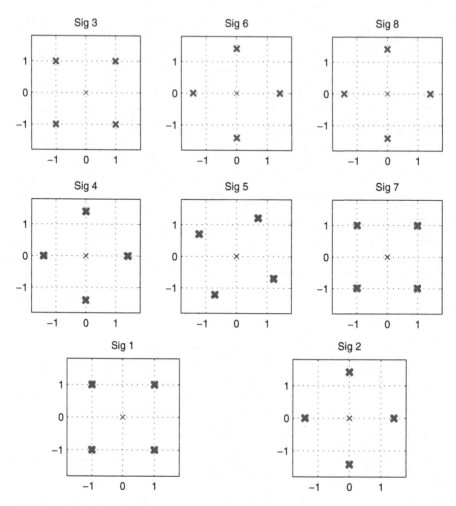

Fig. 3.28 Synthesizer outputs: demodulated and reconstructed constellations

3.5   Recap

In this chapter we introduced basics on multirate signal processing. The standard structure of the polyphase down converter channelizer has also been derived. It represents the key engine in the back-end receiver design presented here. Some good references to better understand the dual up converter channelizer structure that is used in the back-end transmitter design are given at the end of this chapter. They can also be useful for understanding the modifications on the channelizer structure used in both the proposed structures.

The structures proposed here can obviously be used in a software radio; in fact they achieve the fundamental goal of simultaneously up and down converting arbitrary bandwidth signals with randomly located center frequencies.

References

1. fred j. harris, *Multirate Signal Processing for Communication systems*, Prentice Hall, 2004.
2. fred harris, Chris Dick, and Michael Rice, "Digital Receivers and Transmitters Using Polyphase Filter Banks for Wireless Communications," Special Issue of *Microwave Theory and Techniques, MTT*, vol. 51, no. 4, 2003.
3. f. harris and W. Lowdermilk, "Software Defined Radio: a Tutorial" *IEEE Instrumentation and Measurement Magazine*, vol. 13, no.1, 2010.
4. f. harris, C. Dick, X. Chen, and E. Venosa, "Wideband 160 Channel Polyphase Filter Bank Cable TV Channelizer," *IET Signal Processing*, 2010.
5. f. harris, C. Dick, X. Chen, E.Venosa, "M-Path Channelizer with Arbitrary Center Frequency Assignments," *Proc. of WPMC 2010 - 13-th International Symposium on Wireless Personal Multimedia Communications*, Recife, Brazil, October 11-14 2010.
6. E. Venosa, X. Chen, f. harris, "Polyphase Analysis Filter Bank Down-Converts Unequal Channel Bandwidths with Arbitrary Center Frequencies - Design I," *Proc. of SDR'10 - Software Defined Radio Technical Conference and Product Exposition*, Washington DC, November 30-December 3, 2010.
7. X. Chen, E. Venosa, f. harris, "Polyphase Synthesis Filter Bank Up-Converts Unequal Channel Bandwidths with Arbitrary Center Frequencies - Design II," *Proc. of SDR'10 - Software Defined Radio Technical Conference and Product Exposition*, Washington DC, November 30-December 3, 2010.
8. fred harris, "Polyphase Filter Bank for Unequal Channel Bandwidths and Arbitrary Center Frequencies," *Proc. of SDR'10 - Software Defined Radio Technical Conference and Product Exposition*, Washington DC, November 30-December 3, 2010.

Chapter 4
Spectral Analysis

4.1 Introduction

Spectrum analyzers for communication systems are traditionally modeled by a bank of narrowband filters with equally spaced frequency centers and bandwidths. This matches the structure of communication system channelized bandwidth frequency assignments that in turn influenced the design of early swept frequency spectrum analyzers. They were designed around a fixed bandwidth intermediate frequency filter through which spectral regions are probed by shifting them to the filter with a linear time-frequency swept heterodyne.

The spectrum analyzers for communication systems perform a decomposition of the input signal into basis functions that share common features such as time extent and bandwidth.

Moreover, the model of the ear as well of other mechanical resonator systems such as vibrating strings and air columns exhibit bandwidths that are proportional to their center frequencies. Analyzers matched to this spectral characteristic are called proportional bandwidth or constant-Q spectrum analyzers where Q, the quality factor of a filter, is the ratio of a filters center frequency to its bandwidth. Those spectrum analyzers are modeled by a bank of filters with increasing spacing of spectral centers and bandwidths. The filters are seen to have equally spaced centers and equal bandwidths on a logarithmic scale. This type of spectrum analyzer performs a decomposition of the input signal into basis functions that share a scaled range of features such as varied time duration and varied bandwidths.

Analysis of mechanical systems such as vibrating beams, acoustic resonators, cochlea of the human ear, whale and dolphin sounds, harmonic rich FM waveforms, and image features are best performed by proportional bandwidth spectrum analyzers. They can be used in cognitive sensor networks and in many other software defined radio applications.

Note that stationary and non stationary signals require different criteria for the proportional factor coupling bandwidth to center frequency. For stationary audio signals such as speech and music, the bandwidth is proportional to the signal center

E. Venosa et al., *Software Radio: Sampling Rate Selection, Design and Synchronization,*
Analog Circuits and Signal Processing, DOI 10.1007/978-1-4614-0113-1_4,
© Springer Science+Business Media, LLC 2012

frequency so that two center frequencies an octave apart will have filter bandwidths with a ratio of 2. Graphic equalizers and sound boards used for audio recording and playback are constant-Q filter banks with equally spaced centers and equal bandwidth on a log scale. This property is useful for tracking harmonics which move unequal intervals in standard spectrum analyzers but move the same interval in a constant-Q spectrum analyzer. For instance, if a fundamental tone moves 10% of its center frequency, its first harmonic moves 20% and its second harmonic moves 40%.

For non stationary signals the bandwidths are chosen to be proportional to the square root of the center frequency so that two center frequencies an octave apart have filter bandwidths with a ratio of $sqrt(2)$. The square root proportional bandwidth spectrum analyzer offers the maximum integration gain for linearly varying FM sweeps. Investigators of dolphin communication signals, which are linear FM sweeps, use this form of spectrum analyzer.

In this chapter of the book we present three new structures for octave processor of constant-Q spectrum analyzers. All of them share the common characteristic of presenting a very small workload when compared with the traditional spectrum analyzers.

4.2 Constant-Q Spectrum Analyzer Architecture Using Hilbert Transform Filter

The architecture of a digital proportional bandwidth spectrum analyzer consists of a cascade of two parallel processing threads as shown in Fig. 4.1.

The first octave processor partitions the top octave of the input signal bandwidth into the desired number of logarithmically spaced, proportional bandwidth filters. Often the number of filters is 6 to 12 filters per octave though systems with many more partitions have been reported. The second processor performs half-band filtering along with a 2-to-1 down sampling commensurate with the 2-to-1 bandwidth reduction.

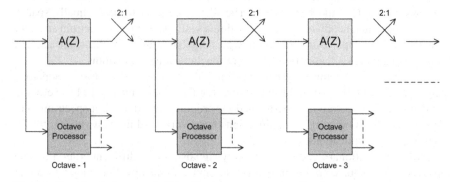

Fig. 4.1 Cascade of concurrent half-band and octave processing filters

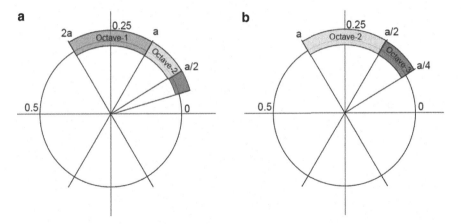

Fig. 4.2 Top octave at sample rate f_s replaced by next octave at sample rate $f_s/2$

As a result of the 2-to-1 down sampling operation the frequency band occupying the next lowest octave becomes the top octave at the reduced sample rate. Thus only the top octave needs to be spanned in the proportional bandwidth filter set while the pre-filtering performed by the half-band filter and 2-to-1 down sampling iteratively delivers subsequent octaves to the same fractional frequency band at successively reduced sample rates. The shifting of consecutive octaves to top octave status in successive half-band, half sample rate processing paths is illustrated in Fig. 4.2.

The frequency denoted a is the lower end of the top octave with the upper end of that octave denoted 2a. The frequencies a and 2a can be any fraction of the sample rate but it is reasonable to have 2a in the circles second quadrant while a is in the circles first quadrant.

To assure a reasonable transition bandwidth for the concurrent half-band filter we would like the frequency corresponding to 2a to be less than or equal to $\frac{3}{8}f_s$. Reasonable contender sets for the boundaries of the top octave span the frequency intervals at the high end from $\frac{3}{16}f_s$-to-$\frac{6}{16}f_s$ or at the low end from $\frac{1}{8}f_s$-to-$\frac{2}{8}f_s$. A particularly good choice is the span from $\frac{1}{6}f_s$-to-$\frac{2}{6}f_s$. As shown shortly, this choice permits a clever signal conditioning sequence of filtering and 2-to-1 down sampling as part of the octave filter processing.

4.2.1 Octave Processing Block

For the proportional bandwidth analyzer design presented here, the boundaries of the top octave were selected to span the interval from $\frac{1}{6}$-to-$\frac{2}{6}$ of the input sample rate. We will use a specific input sample rate to more easily refer to specific frequencies and frequency spans. If we assume an input sample rate of 120 kHz, then the top octave spans the interval between 20 kHz to 40 kHz. We chose this span

Fig. 4.3 Pre-processing 2-to-1 down sample blocks of octave filter bank

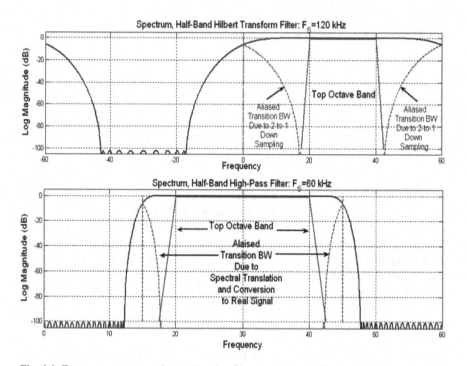

Fig. 4.4 Frequency responses of pre-processing filters

to be symmetric about the quarter sample rate of 30 kHz. We will have easy access to this band following subsequent half-band filtering and 2-to-1 down sampling. The purpose of this 2-to-1 down sampling is to reduce the computational burden on the proportional filter bank in this block and should not be confused with the 2-to-1 down sampling in the concurrent parallel block that shifts successive octaves into the top octave position.

The pre-processing blocks that implement the 2-to-1 down sampling in the octave filter block are shown in Fig. 4.3.

This process uses two half-band filters to reduce the signal bandwidth by a factor of 4 but only reduces the sample rate by a factor of 2-to-1. The frequency responses of the two filters implemented as FIR filters are shown in Fig. 4.4 while Figs. 4.5 and 4.6 illustrate the spectral transformations performed by the pre-processor.

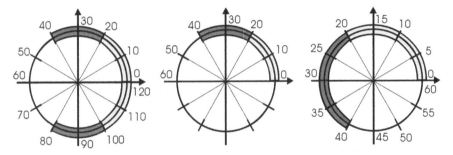

Fig. 4.5 Spectra at input and output of 2-to-1 down sampled Hilbert transform filter

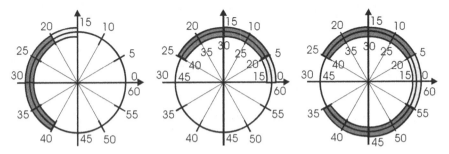

Fig. 4.6 Spectra at input and output of half-band high-pass filter, quarter sample rate heterodyne and complex to real conversion

The pre-processor first applies a Hilbert transform filter to pass the positive frequency spectral segment of the input signal shown on the left side of Fig. 4.5 to obtain the spectrum shown in the center of Fig. 4.5. The filtered signal is down sampled 2-to-1 to alias the top octave band originally centered at the quarter sample rate of the input sample rate to the half sample rate of the output sample rate. The aliased spectrum is shown in the right side of Fig. 4.5.

The Hilbert transform and the 2-to-1 down sampling are actually performed as a single combined operation but have been described as separate operations for clarity of description of the spectral transformations.

As shown in the left side of Fig. 4.6, the down sampled time series is then filtered by a half-band high-pass filter to restrict the bandwidth to the spectral span extending from $\frac{1}{8}f_s$ to $\frac{3}{8}f_s$, i.e., from 15 to 45 kHz in the first pass processing block. This span contains the desired top octave extending from $\frac{1}{6}f_s$ to $\frac{2}{6}f_s$, i.e., from 20 to 40 kHz in the first pass processing block. The spectrum with this freshly restricted bandwidth centered at the half sample rate is then heterodyned back to the quarter sample rate to obtain the spectrum shown in the center of Fig. 4.6. The shifted signal, currently represented by complex samples, is converted back to a real signal by discarding the imaginary samples to obtain the Hermitian symmetric

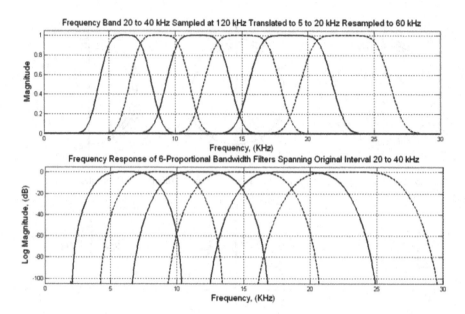

Fig. 4.7 Frequency responses of six proportional bandwidth filters

spectrum shown in the right side of Fig. 4.6. In fact, only the real samples are formed in the heterodyne process but again we separated the steps for clarity of description.

Note here that the spectrum of the top octave band originally occupying the 20 kHz span between 20-to-40 kHz at the 120 kHz input sample rate now occupies the 20 kHz span between 5-to-25 kHz at the 60 kHz output sample rate.

The real signal samples formed by the pre-processing blocks are then passed on for processing by the proportional bandwidth filter bank. Converting the filtered, down sampled and spectrally centered top octave to a real signal reduces the computational work load in the following proportional bandwidth filter bank.

4.2.2   Proportional Bandwidth Filters Design

Figure 4.7 shows the magnitude and log-magnitude frequency responses of 6 proportional bandwidth filters implemented as FIR filters.

As described earlier, the frequency band they span is the top octave of the input signal extending from 20-to-40 kHz for the input sample rate of 120 kHz. The preprocessing has placed this same band in the frequency interval 5-to-25 kHz at the output sample rate of 60 kHz. The proportional bandwidth filters operate within the offset frequency band at the reduced rate and at reduced implementation cost. The energy extracted from each proportional frequency is presented in spectral displays at the center frequency from which the signal was translated. This of course is a 25 kHz offset from the center frequency of each of the proportional bandwidth filters.

We note that the widest bandwidth filter, the rightmost of the filters in the octave filter bank has a 3 dB and a 100 dB bandwidth of approximately 6 kHz and 13.5 kHz respectively. Since the input sample rate to the filter bank is 60 kHz, the reduced bandwidth output time series from each filter in the octave filter bank is oversampled and can be down sampled as a part of the filtering operation. This down sample ratio can be as small as 4-to-1 to avoid aliasing of the 100 dB band edges of the widest filter or can be as large as 10-to-1 to match the 3 dB bandwidth.

A reasonable compromise would be 6-to-1 down sampling for an output sample rate of 10 kHz corresponding to the 40 dB bandwidth of the widest bandwidth filter in the bank. In the design presented here, the number of taps or coefficients required to implement each FIR filter in the filter bank is 90. We can operate the filter bank at a 10 kHz output rate rather than its 60 kHz input rate. In doing so we distribute the workload per output sample over six input samples, converting the 90 operations per output sample to 15 operations per input sample. For ease of memory management we can operate all six filters in the filter bank at the same output rate.

4.2.3   Sample Rate Reduction Half-Band Filter

The remaining filter required to implement the proportional bandwidth filter bank is the half-band filter that reduces the input bandwidth by a factor of 2 while simultaneously reducing the sample rate by the same 2-to-1 ratio. The two pre-processing filters used in the octave band processing block were also half-band filters with one reducing bandwidth and sample rate and the other simply reducing bandwidth. The attraction of these half-band filters is that, when properly designed, alternate samples are zero and the non-zero samples are even symmetric. When the FIR filter is implemented in the folded form, the N-tap filter requires only $\frac{N}{4}$ multiplies per output sample point. If the filter also supports the 2-to-1 down sample operation, the workload further drops to $\frac{N}{8}$ operations per input sample point. Continuing this reasoning, the half-band 2-to-1 resampling filter responsible for rolling successive octaves into the top octave position as successively lower sample rates is also implemented as an exact half-band filter.

Figure 4.8 shows the impulse and frequency response of this 47-tap half-band filter. Also indicated here, as was shown in the earlier versions of half-band filter, is the signal bandwidth being extracted by the filter as well as the aliased transition band edges of the filter caused by the 2-to-1 down sampling operation. We see that the aliased spectrum is suppressed to 100 dB by the filter stop band response prior to being folded into the desired filter pass-band bandwidth.

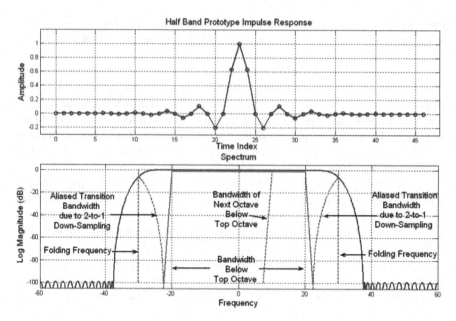

Fig. 4.8 Impulse response and frequency response of half-band 2-to-1 down sampling FIR filter

4.2.4 Simulation Results

The frequency domain performance of the iterative octave processor and half-band processor can be seen in Fig. 4.9.

The upper subplot shows the frequency response of the half-band 2-to-1 down sampling filter at four successive half sample rates displayed on the frequency axis of the input signal sample rate. These successive filters are responsible for the bandwidth reduction and sample rate change that slides successive octaves of the input signals spectrum to occupy the top octave position and participate in the proportional bandwidth partition.

Earlier we presented the frequency response of the single top octave processed at the translated and down sampled frequency. The lower subplot of Fig. 4.9 shows the equivalent up sampled and frequency offset spectra of the same top octave plotted in the band from which the octave had been translated. The subplot further shows the spectra of the partitioned bands in the next two lower octaves. In this figure the right most positioned filter in the top octave is centered at approximately 37.4 kHz with a 3 kHz bandwidth while the left most positioned filter in the bottom octave is centered at approximately 3.1 kHz with a 0.82 kHz bandwidth. We would require 66 filters to span the input frequency band extending from 20 Hz to 40 kHz. We would also require a logarithmic display to see the spectral response over the 11 octave span.

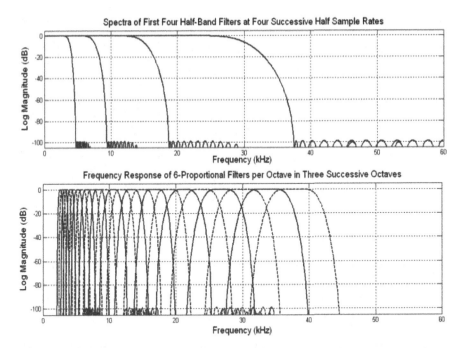

Fig. 4.9 Spectra of half-band 2-to-1 resampling filter at four successively lower half sample rates and the spectra of six proportional bandwidth filters displayed at correct offset frequency for three successive octaves

4.2.5   *Workload Analysis*

We now examine the computational workload requirements of the proportional bandwidth spectrum analyzer. We have commented that particular filters have been selected on the basis of their computationally efficiency. We now address that efficiency.

Our first task is to estimate the workload of the octave processor composed of the two half-band filters and the six proportional bandwidth filters. All filters were designed to support the near 100 dB dynamic range of a 16-bit analog to digital converter. The result of this analysis task is compactly tabulated in Table 4.1.

We start with the pre-processor Hilbert transform filter that was designed as an exact half-band FIR filter of length 21 taps with 10 non-zero coefficients. When the 2-to-1 down sample is embedded in this filter, the workload is five multiplies per input sample. We will ignore the possible workload reduction due to coefficient symmetry. The second pre-processor filter is the half-band filter designed as an exact half-band FIR filter of length 67 taps with 34 non-zero coefficients. There is one each of this filter operating on the half rate real and imaginary components of the Hilbert transform filter output. Referenced back to the input rate, each filter requires seventeen multiplies per input sample. Each of the six proportional bandwidth filters

Table 4.1 Filter type, length, resample ratio, sample rate and multiplies per input sample

Sample Rate and Multiplies per Input Sample

Filter	# taps	# Non Zero	M:1	F_s (kHZ)	Mult Per Input
Hilbert	21	10	2:1	120	5
HB-RL	67	34	1:1	60	17
HB-IM	67	34	1:1	60	17
P BW-1	90	90	6:1	60	7.5
P BW-2	90	90	6:1	60	7.5
P BW-3	90	90	6:1	60	7.5
P BW-4	90	90	6:1	60	7.5
P BW-5	90	90	6:1	60	7.5
P BW-6	90	90	6:1	60	7.5
HB 2:1	47	24	2:1	120	12
Total	674	-	-	-	96

were designed as a FIR filter for one of the octave bandwidths spanning the octave and each required 90 taps. We commented earlier that these filters could be down sampled 6-to-1 and when referenced to the input rate, the workload per filter is 7.5 multiplies per input sample.

Finally the half-band 2-to-1 down sample filter was designed as an exact half-band FIR filter of length 47 taps with 24 non-zero coefficients. When the 2-to-1 down sample is embedded in this filter, the workload is twelve multiplies per input sample. Summing the workload for all the filters processing the top octave we find the total workload to be 96 multiplies per input sample to implement the top octave. Each successive octave operates at half the clock rate with half the workload per successive octave.

The total workload for a large number of octaves is the sequence

$$96 * (1 + 1/2 + 1/4 + 1/8 + \cdots) \tag{4.1}$$

which is bounded by 192 multiplies per input sample. It is remarkable that the computational requirement to compute the output time series from all the octave filters is less than 192 multiplies per input sample.

4.3 Constant-Q Spectrum Analyzer Architecture Using Polyphase Filter Bank

The constant-Q spectrum analyzer presented in this section uses only a 4-path polyphase channelizer along with a proportional bandwidth filter bank. The 4-path polyphase channelizer is used to perform 4-to-2 down sampling. Remember that this engine is able to simultaneously shift lower octaves to the same spectral interval and translate the current octave to base-band allowing the proportional filter bank to work on a reduced sampling rate and decreasing the total workload of the system. To further reduce the workload recursive filters are used to implement the proportional bandwidth filters.

4.3.1 Proposed Architecture

In Fig. 4.1 we showed the block diagram of a conventional constant-Q spectrum analyzer. As we said in the previous section, because of the half-band filtering and the 2-to-1 down sampling, the frequency band occupying the next lowest octave slides up to occupy the spectral interval of the top octave at the reduced sample rate. Thus, a single octave processing scheme is applied to the successively lower octaves. By applying this process ten times we can cover the entire frequency range from 20 Hz to 2000 Hz.

In the standard spectrum analyzers the sub-octave processor directly decomposes the selected spectrum using a bank of proportional bandwidth filters. This bank is usually composed of 3, 6 or 12 filters with the lower frequency of one filter corresponding to the upper frequency of the adjacent filter.

We recall here that the structure presented in Fig. 4.1 has an interesting property: if we assume that the processing workload of the first stage, the input stage composed of the top octave processor and of the first half-band filter, is L operations per input sample, the next stage requires the same processing workload but, operating at half clock rate, it requires $\frac{L}{2}$ operations per input sample. It is simple to verify that the workload for the complete sequence of stages results in

$$\frac{Ops}{Input} = L + \frac{L}{2} + \frac{L}{4} + \frac{L}{8} + \cdots \qquad (4.2)$$

From (4.2) we can see that the workload for an arbitrary large number of stages is always less than twice the workload of the first stage.

It would be easy to reduce the workload of such a structure by applying a 4-to-1 down sampling to the current octave before processing it. We could use a 4-path polyphase channelizer that has the effect of shifting the signal to base-band but this will be a huge wastage of engineering resources. We can in fact use this powerful structure in a much better way obtaining much more than a simple 4-to-1 down sampling of the input octaves.

Figure 4.10 shows the novel constant-Q spectrum analyzer, it is composed only of a 4-path polyphase down converter performing 4-to-2 down sampling and of an octave processor which is composed of a bank of constant-Q filters.

Figure 4.11 shows the effects of the channelization process in the frequency domain for one of the stages composing the spectrum analyzer, here it is clearly shown that the 4-path channelizer has the double effect of shifting the current octave to base band (Fig. 4.11c), and simultaneously translating the next octave in the top octave spectral range (Fig. 4.11b).

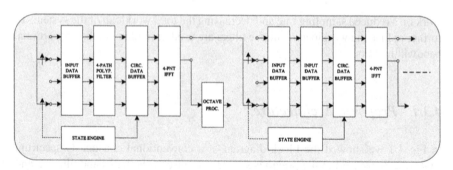

Fig. 4.10 Block scheme of the proposed architecture

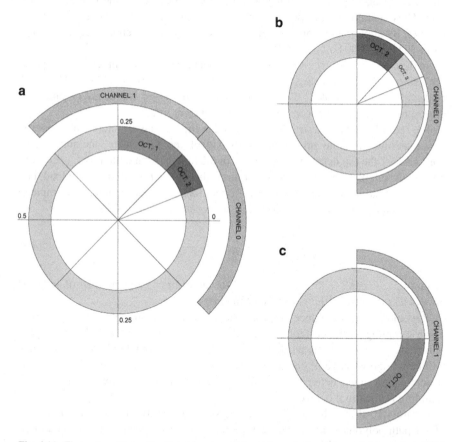

Fig. 4.11 First octave channelization process and first (channel 0) and second (channel 1) IFFT outputs

Note that in our design the top octave is located between $\frac{1}{8}$ and $\frac{1}{4}$ in the normalized frequency domain. We are also assuming that the two-side bandwidth of the sampled input signal lies in the frequency interval $[-0.25, 0.25]$. Obviously at the output of

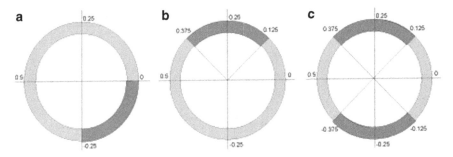

Fig. 4.12 First octave processing: base-band complex octave, shifted complex octave, and shifted real octave

4-PNT IFFT block we achieve four outputs. But, for our proposal, we simply discard the last two considering only the outputs number one (channel 0) and number two (channel 1) because they are the only one containing the signal that needs to be processed.

As it is shown in Fig. 4.11b, c, the octave translated to base band is a complex signal, lying in the frequency interval $[-0.25, 0]$; we need to shift it before processing. A free $90°$ shifting translates the signal in the frequency range $[0, 0.25]$ but, because recursive filters are used to design the proportional bandwidth filter bank, if we process the octave directly in this spectral interval, for the first filter, the one with the lower edge equal to zero, we encounter the problem due to the spectral folding. In order to avoid this, before processing the octave we need to shift it elsewhere. We could think of shifting the octave by $180°$ because it still doesn't cost any multiplies, but this is not a good option, in fact, by doing so we find the octave in $[0.25, 0.5]$ encountering again the same spectral folding problem around the half sampling rate frequency for the largest constant-Q filter. The shift we decide to perform here is $135°$ so that we can later filter the signal in the frequency interval $[0.125, 0.375]$. This process is described in Fig. 4.12b.

Before processing the octave in the proportional bandwidth filter bank we discard its imaginary part making it real and halving the number of multiplications to be performed. As shown in Fig. 4.12c this operation has the effect of doubling the signal bandwidth.

4.3.2   Simulation Results

In this section we show the simulation results for the proposed spectrum analyzer.

In particular, Fig. 4.13 shows the frequency response of a bank of six proportional bandwidth filters composing the octave processor. They span the frequency range from 0.125 to 0.25 in the normalized domain. This is the frequency location of the top octave.

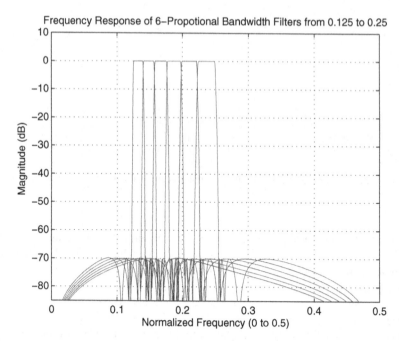

Fig. 4.13 Frequency responses of six constant-Q filters in the frequency interval $[0.125, 0.25]$ in the normalized frequency domain

The simulation shows clearly that these filters present a constant-Q factor in the linear scale while they are equally spaced and have equal bandwidths on a log scale. As a consequence of the 4-to-2 down conversion and frequency shifting processing due to the 4-path channelizer and also because of the 135° frequency translation, their original frequency band is translated to the frequency interval $[0.125, 0.375]$.

The shifted filters are shown in Fig. 4.14. These filters are all 9th order recursive filters designed by using all-pass networks. The Matlab routine *tony_des_2* has been used to generate the filter's coefficients. Each of the filter is composed of only 18 coefficients.

Figure 4.15 presents the channel 0 (first IFFT output) frequency response of the 4-path down converter channelizer for four successive stages in the normalized frequency domain. This figure clearly shows the bandwidth and sample rate reduction due to the down converter channelizer at any stage of the processing chain.

The complete frequency domain performance of the 4-stage spectrum analyzer is shown in Fig. 4.16. The upper plot shows the behavior of the proportional bandwidth filters chain on the logarithmic scale. As expected, they are equally spaced on the full spanned frequency range. The lower plot displays the same filters on the linear scale, here the filter bandwidths increase when their center frequencies increase.

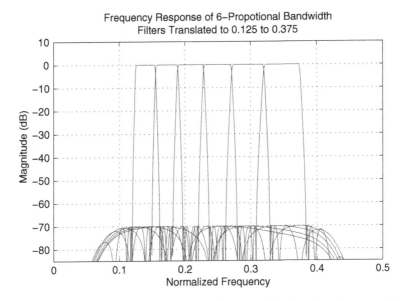

Fig. 4.14 Frequency responses of six constant-Q filters shifted to the frequency interval $[0.125, 0.375]$ in the normalized frequency domain

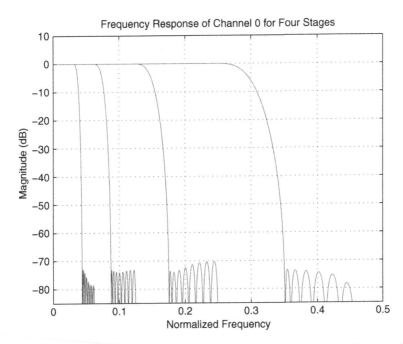

Fig. 4.15 Channel 0 frequency response of the 4-path down converter channelizer for four successive stages in the normalized frequency domain

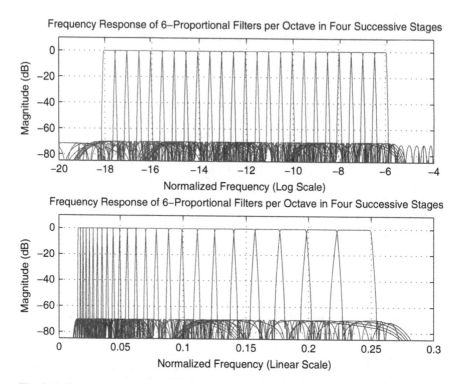

Fig. 4.16 Frequency domain performance of 4-stage spectrum analyzer structure

4.3.3 Workload Analysis

In this section we examine the computational workload requirements of the novel proportional bandwidth spectrum analyzer. We start by confining our scope within one single processing stage.

The low-pass prototype filter used in the 4-path down converter channelizer has 79 taps and it requires 79 multiplies to compute one output for every two inputs. Before being processed in the proportional filter bank, the signal experiences a $\frac{3\pi}{4}$ frequency shift, which is a complex multiplier. It requires 4 multiplies for each input sample. After frequency shifting, we also discard the imaginary part of the signal before processing it in the proportional bandwidth filter bank. We consider this filter bank composed of six proportional bandwidth filters. Since each of the filter has 18 coefficients, the bank will take 108 multiplies to produce one output for each octave.

Summarizing, the first stage of the proposed spectrum analyzer performs $79 + 4 + 108 = 191$ operations to produce one output. In other words, the total workload for the first stage is $N = 191/2 = 95.5$ multiplies per input sample.

The next stage requires the same workload but operating at half sampling rate it requires $\frac{N}{2}$ operations per input sample. In the same way the third stage requires

$\frac{N}{4}$ operations while the fourth one requires $\frac{N}{8}$ operations. The total workload of the structure is

$$95.5 + 48 + 24 + 12 = 179.5. \tag{4.3}$$

We conclude this section specifying that for an arbitrary number of stages the total workload of the proposed spectrum analyzer is

$$\frac{Ops}{Input} = 95.5 + \frac{95.5}{2} + \frac{95.5}{4} + \frac{95.5}{8} + \cdots < 191. \tag{4.4}$$

4.4 Constant-Q Spectrum Analyzer Architecture Using FFT

In this section we present another extremely efficient FFT based spectrum analyzer that performs constant-Q spectral analysis. Fourier transforms are normally used for performing equally spaced, equal bandwidth spectral estimation; their main advantage is the efficiency of the FFT algorithm.

By post processing the output of the FFT block with a frequency sliding window having variable bandwidth, we are able to achieve spectral estimations at arbitrary center frequencies with arbitrary resolutions maintaining the advantages of using the FFT algorithm.

To further decrease the total workload, a 4-path polyphase channelizer is used to down sample the octaves before processing them in the FFT block.

We recall that, in the spectral analyzer depicted in Fig. 4.1 because of the half-band filtering and the successive 2-to-1 down sampling, the frequency band occupying the next lowest octave slides up to occupy the spectral interval of the top octave at the reduced sample rate. By using this approach, only one spectral interval, the same for every stage, needs to be processed by the octave processor. The 2-to-1 down sampling process is depicted in Fig. 4.2. In particular, Fig. 4.2a, b respectively show the normalized signal spectrum before and after down sampling.

In such a structure, the octave processor can be composed of an FFT block that directly processes the selected spectrum. It is usually followed by a frequency sliding window. The use of FFT algorithm should make this structure extremely efficient but, by taking into account the pre and the post processing stages in the workload calculation, we observe the following: to begin with, as we are processing the top octave, the number of FFT points in every stage has to be very large. It directly affects the resolution of the spectrum analyzer. Also, in the proportional bandwidth filtering stage, only 25% of the FFT output sample points are used. We could use an appropriate pruning of the FFT by computing only the positive frequencies in the desired octave but still we have to consider the windowing stage. It applies a time domain window as a circular convolution in the frequency domain by sliding a position-dependent set of weights through the FFT bins to form the desired weighted sum. The sliding weights are obtained from the main lobe of the Fourier transform of a time domain window.

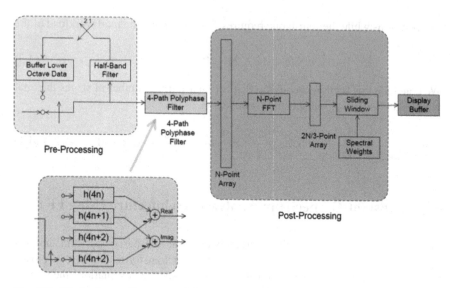

Fig. 4.17 Block scheme of one stage of the proposed spectrum analyzer

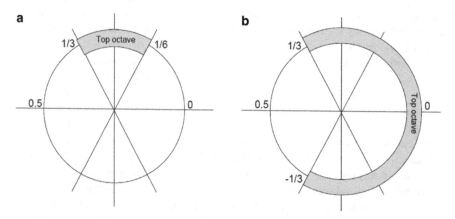

Fig. 4.18 4-to-1 down conversion processing; (**a**) top octave before down sampling, (**b**) top octave after 4-to-1 down sampling

It is clear that the number of FFT points along with the way in which the output samples are used in the following filtering stage affects the total workload of such a system. Also a wise choice of the sliding window is an important factor.

In order to decrease the total workload we need to reduce the sampling rate of the octaves before passing them through the FFT block. We achieve this goal by inserting a 4-path polyphase down converter in each of the stages of the conventional spectrum analyzer.

One of the stages of the proposed structure is shown in Fig. 4.17 while Fig. 4.18 shows the channelizer effects on the top octave spectrum in the normalized frequency domain. This plot clearly shows that the 4-path channelizer has the double

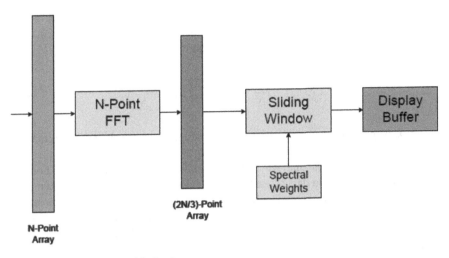

Fig. 4.19 Octave processor block scheme

effect of shifting the current octave to base band (Fig. 4.18b) and of reducing its sample rate.

Note that to ensure a reasonable transition bandwidth for the half-band filters we consider the highest normalized frequency of the top octave to be less than or equal to $\frac{3}{8}$. In addition, to ensure a reasonable transition bandwidth for the 4-to-1 down converters we have to consider the spanned frequency range to be within $\frac{1}{3}$ and $\frac{1}{6}$. This last assumption prevents us from using a down sampling factor larger than four.

4.4.1 Octave Processor Design

The octave processor used in the proposed spectrum analyzer is shown in Fig. 4.19. It is formed of an N-point FFT block followed by a sliding window that obtains the spectral estimate from the proportional bandwidth filters.

The filter bank obtained from FFT is characterized by equally spaced spectral centers and equal bandwidths spanning specific segments of the frequency axis. Our goal is to achieve the spectral estimate at arbitrary center frequencies. Also the filter bandwidths have to change proportionally with their center frequency. We achieve the arbitrary center frequency spectral estimation by using (4.5)

$$F^w(\theta) = \frac{1}{N} \sum_{k=0}^{N-1} F_k W \left(\theta - k\frac{2\pi}{N} \right) \tag{4.5}$$

where $F_k = F[k\frac{2\pi}{N}]$ and $W_k = W[k\frac{2\pi}{N}]$ are the discrete Fourier transform of the signal f_n and of the window w_n respectively.

Equation 4.5 is the N-point discrete Fourier transform of the windowed signal $(f_n \cdot w_n)$ achieved by substituting f_n with its inverse Fourier transform and interchanging the order of summation. Evaluating it at θ_0 we achieve (4.6)

$$F^w(\theta_0) = \frac{1}{N} \sum_{k=0}^{N-1} F_k W \left(\theta_0 - k\frac{2\pi}{N} \right) \tag{4.6}$$

where θ_0 is the arbitrary center frequency at which we want to achieve the estimation and it does not necessarily coincide with multiples of $\frac{2\pi}{N}$.

Equation (4.6) states that we have, via FFT, N samples F_k of the Fourier transform of the N-point data sequence; we also locate the transform of the window $W[k\frac{2\pi}{N}]$ at the desired sample location θ_0 and compute its samples at the locations corresponding to multiples of $\frac{2\pi}{N}$. Then, as defined, we calculate the summation of the products scaled by $1/N$ achieving the estimate of the Fourier spectrum at the location θ_0.

Note that this processing simply applies a time domain window as a convolution in the frequency domain rather than as a multiplication in the time domain. The convolution is accomplished by sliding a position dependent set of weights through the FFT bins to form the position dependent weighted sum. The weights are obtained from the main lobe of the transform of the minimum 4-term Blackman-Harris window. It is defined by (4.7).

$$w(n) = a_0 - a_1 cos\left(\frac{2\pi}{N}n\right) + a_2 cos\left(\frac{2\pi}{N}2n\right)$$

$$-a_3 cos\left(\frac{2\pi}{N}3n\right), n = 0, \dots, N-1. \tag{4.7}$$

The parameters a_i with $i = 0, 1, 2, 3$ are listed in [7]. The impulse response of this window along with its normalized frequency response for a 128-point FFT is given in Fig. 4.20.

An example of bandwidth tuning is given in Fig. 4.21. This plot shows eleven bandwidths obtained by resampling the main lobe of the Fourier transform of a Blackman-Harris window.

4.4.2 Simulation Results

In this section we demonstrate the performance of the proposed constant-Q spectrum analyzer via simulations.

In particular, Fig. 4.22 shows the impulse response and the frequency response of the half-band filter connecting each stage. It is a 33 taps long filter; its pass band goes from 0 to $\frac{1}{6}$ on the normalized frequency axis while its stop band is from $\frac{2}{6}$ to $\frac{1}{2}$. As shown in this figure, these ranges guarantee no spectral folding in the pass band of the filter after the 2-to-1 down sampling.

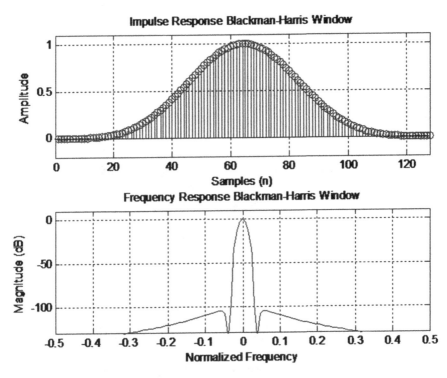

Fig. 4.20 Impulse response and frequency response of 4-term Blackman-Harris window

Figure 4.23 depicts the impulse response and the frequency response of the low pass prototype filter for the 4-path polyphase down converter. This filter is 135 taps long. Its pass-band goes from 0 to $\frac{1}{12}$ on the normalized frequency axis while its stop band is from $\frac{1}{8}$ to $\frac{1}{2}$. Both of these filters are Blackman-Harris windows exhibiting $-92\,dB$ stop band performance.

We have tested the performance of the proposed structure with a signal composed of ten sinusoids with different center frequencies. The normalized center frequencies of these sinusoids are 0.0078, 0.0117, 0.0172, 0.0219, 0.0313, 0.0469, 0.0781, 0.1094, 0.1406, 0.2188. The normalized spectrum of this signal is shown in the upper subplot of Fig. 4.24. This plot is obtained by first windowing the time series of the input signal with a 4-term Blackman-Harris window and then taking a 1024-point FFT. It should be noted that this processing is not able to perfectly resolve some of the sinusoids whose center frequencies are very close to each other, i.e. the sinusoids centered on 0.0078 and 0.0117 can hardly be distinguished.

The lower subplot of Fig. 4.24 demonstrates the output of the proposed constant-Q spectrum analyzer. We let the test signal go through five stages of the structure, thus the analyzed frequency range at the output is, in the normalized frequency domain, from $\frac{1}{96}$ to $\frac{1}{3}$.

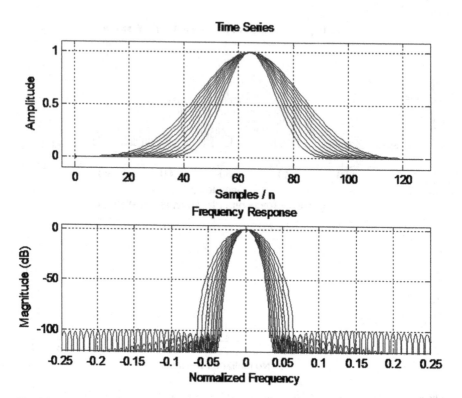

Fig. 4.21 Impulse response and frequency response of 4-term Blackman-Harris window with variable bandwidths

4.4.3   Workload Analysis

For our purposes we initially confine the workload calculation to only one stage and later, by applying (4.10) we extend it to the complete structure.

The half-band filter is a 33 taps long and it requires 33 multiplies to compute one output for every two input samples. The low-pass prototype filter used in the 4-path down converter channelizer has 135 taps and it requires 135 multiplies to compute one output for every four input samples. Before being processed by the filter bank, the input signal is delivered to the 128-point FFT. The workload of this block is $2N log_2(N) = 1792$. We considered the filter bank composed of sixteen proportional bandwidth filters.

Note that the transform based spectral decomposition is not efficient for small number of filters per octave, and that number is in the neighborhood of 16 filters. When we have less than 16 filters in an octave we should implement them as parallel filters.

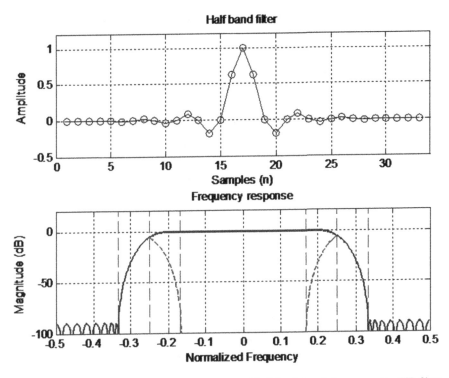

Fig. 4.22 Impulse response and frequency response of half-band 2-to-1 down sampling FIR filter

The length of the filter taps used to achieve the final values go from 8 to 15 so we can consider the average number of filter taps to be equal to 11. Summarizing, the first stage of the proposed spectrum analyzer performs a number of operation that is indicated in (4.8).

$$L = \frac{33}{2} + \frac{135}{4} + \frac{1792}{128} + \frac{11 * 16}{128} = 65.625. \tag{4.8}$$

The second stage requires the same workload but operating at half sampling rate it requires $\frac{L}{2}$ multiplies per input sample. In the same way the third stage requires $\frac{L}{4}$ multiplies while the fourth one requires $\frac{L}{8}$ multiplies. By following this reasoning the total workload with a five stage spectrum analyzer can be approximated to

$$65.6 + 32 + 16 + 8 + 4 = 125.6 \tag{4.9}$$

multiplies per input sample.

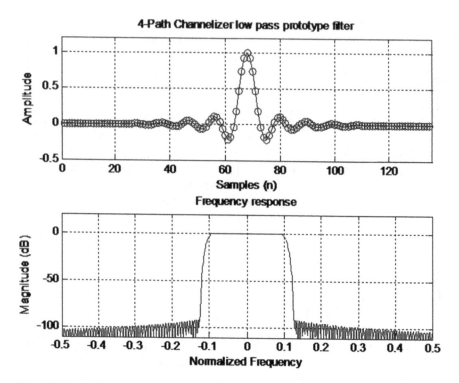

Fig. 4.23 Impulse response and frequency response of low-pass prototype filter of the 4-path polyphase down converter

We conclude this section specifying that for an arbitrary number of stages the total workload of the proposed structure is

$$\frac{Ops}{Input} = 65.625 \left(1 + \frac{1}{2} + \frac{1}{4} + \frac{1}{8} + \cdots \right) < 131.25. \qquad (4.10)$$

4.5 Recap

In this chapter we have presented the design process of three novel and extremely efficient octave processors for constant-Q spectrum analyzer. They could find wide application in acoustics, speech, and vibration analysis. The spectrum analyzers we presented are based on stages. At every stage the workload is halved so that their total workload cannot be more that twice the workload of the first stage. Simulations have also been presented for supporting the theoretical results.

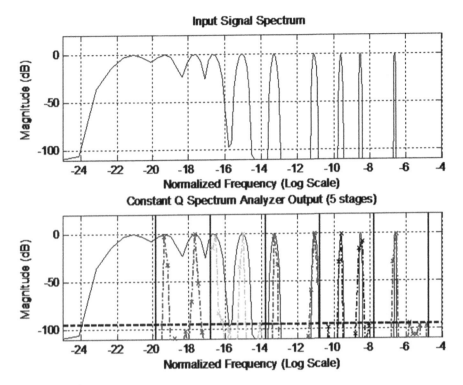

Fig. 4.24 Five stages constant-Q spectrum analyzer output

References

1. fred j. harris, *Multirate Signal Processing for Communication Systems*, Prentice-Hall, 2004.
2. Charles R. Greene, "Proportional Bandwidth Filtering", *IEEE Transactions on Audio and Electroacoustics*, vol. 21, no. 4, 1973.
3. ANSI S1.11-1986 (R1988) American National Standard Specification for Octave-Band and Fractional-Octave-Band Analog and Digital Filters - SAME AS ASA 65-1986
4. Jonathan D. Locke and Paul R White, "Detection Performance of the Fractional Fourier Transform (Chirp FFT) for Frequency Modulated Signals", $3-rd$ *International Conference on Underwater Acoustic Measurements: Technologies and Results*, Nafplion, Greece, 21-26 June 2009.
5. fred j. harris, "An Efficient Constant-Q Spectral Analyzer Architecture Using All-Pass Recursive Filters" Report, Acoustic Analysis Workbench Project, 1999-2000 SPAWAR contract number N66001-97-D-5028, delivery order 0044. Report available on request.
6. fred j. harris, "High-Resolution Spectral Analysis with Arbitrary Spectral Centers and Arbitrary Spectral Resolutions", *Computers and Elect. Eng.*, Pergamon Press, vol. 3, 1976.
7. Fredric j. Harris, "On the Use of Windows for Harmonic Analysis with the Discrete Fourier Transform", *Proceedings of the IEEE*, vol. 66, no.1, 1978.

Chapter 5
Synchronization

5.1 Introduction

In this chapter we consider the time, frequency and phase synchronization issue. We cannot avoid dealing with this topic because no radio receiver can operate without being properly synchronized. If the radio is not synchronized none of the sub-systems composing it can operate: not the matched filters, not the equalizer, not the detectors; not even the error correcting codes or the source decoding will work.

At the beginning of this chapter the topic is introduced and some general notions about synchronization with reference to the physical layer of an actual digital receiver are given. Later in the chapter, a novel algorithm for time, frequency and phase synchronization in an ideal software radio receiver is provided. The proposed receiver model is supposed to operate on QAM signals that have been transmitted through a channel which only adds white Gaussian noise. The receiver operates without analog intermediate frequency down conversion and without splitting the signals into quadrature and in-phase components. Not even the digital down conversion is performed in it. The symbol recovery task is completely assigned to the proposed algorithm running on a receiver that directly converts the received signals into a single stream of values with the help of a single analog to digital converter.

In such a receiver the peculiar synchronization problem arises because of the limited control that the numerical algorithm has on the ADC; it emerges from total asynchronism of symbol rate, carrier frequency and phase.

The proposed algorithm solves the synchronization problem blindly with an information-theoretic based criterion: joint entropy maximization is utilized for frequency and symbol synchronization while mutual information minimization is utilized for phase recovery.

Of course the proposed algorithm cannot be immediately applied to a real digital receiver but it is definitely a good starting point for those people that would like to be introduced to the research on the synchronization issues for a software radio.

E. Venosa et al., *Software Radio: Sampling Rate Selection, Design and Synchronization*, Analog Circuits and Signal Processing, DOI 10.1007/978-1-4614-0113-1_5,
© Springer Science+Business Media, LLC 2012

5.2 Basics of Synchronization

In Fig. 5.1 the hardware model of an actual transmitter-receiver chain is shown. This diagram emphasizes the signal conditioning and signal processing of communication waveforms exiting the modulator and entering the demodulator.

The transmitter of the physical layer model explicitly shows the shaping filter, the up converter, and the output power amplifier which perform the task of baseband spectral shaping, the linear radio frequency spectral transformations, and the non-linear spectral transformations respectively.

We note the asymmetry of the transmitter and the receiver. The receiver, in fact, contains many more subsystems than does the transmitter. These subsystems are seen to be servo control loops that participate in the synchronization and in the signal conditioning tasks required to demodulate the input waveform. These loops estimate the unknown parameters of the input signal and invoke corrective signal processing and signal conditioning operations to ameliorate the degrading effects due to the hardware devices on the demodulation process.

A partial list of these loops is:

1. An automatic gain control (AGC) loop to estimate and remove the unknown channel attenuation;
2. A carrier recovery loop to estimate and remove the unknown frequency offset between the input signals' nominal and actual carrier frequency;
3. A timing recovery loop to estimate and remove unknown time offsets between the receiver sampling clock and the optimal sample positions of the matched filter output series;

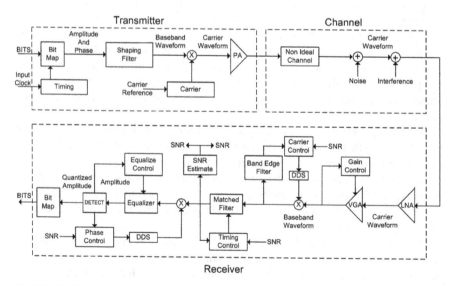

Fig. 5.1 Physical layer model of an actual digital transmitter and receiver

4. An equalizer loop to estimate and remove unknown channel distortion responsible for inter-symbol interference;
5. A phase recovery loop to estimate and remove unknown carrier phase offset between the input signal and the local oscillator;
6. A signal-to-noise ratio estimator to supply important side information to the previously enumerated subsystems.

In the following sections we propose a brief overview of some of the just enumerated loops. In particular we describe the loops used for recovering the timing, phase and frequency of the received signals.

5.2.1 Timing Synchronization

We mentioned in the previous section that the function of the various loops in the receiver is to estimate the various unknown parameters of the received noisy signal. The estimate $\hat{\theta}$, of the phase of the received signal, is based on observations of a sampled data sequence derived from the output of a matched filter that acts to reduce the effects of the additive noise (see Fig. 5.1).

The desired estimate can be obtained from a bank of matched filters parameterized over the unknown variable such a time delay τ. The outputs of the filter bank at specific symbol time increments nT are subjected to a detector and are averaged to obtain stable statistics. The smoothed outputs are compared and the filter with the largest output magnitude is the one matched to the signal time delay τ_k.

In modern receivers, the filter bank is available to the receiver as the paths of an M-path polyphase filter. Rather than operate all the paths simultaneously, they are operated sequentially in response to side information which guides a state machine to the peak of the correlation function. This side information is the slope at the output of each hypothesized filter selection. The system selects any filter in the bank and tests the hypothesis if this it is the correct filter. It does that by forming the derivative at the test point.

In legacy receivers the derivative was estimated from early and late matched filters bracketing the test point in question. In modern receivers it is formed by a derivative matched filter bank. The derivative at selected points of a correlation function is shown in Fig. 5.2 for positive valued and for negative valued correlations. Note that for a positive valued correlation a positive slope indicates the peak is ahead of the test point and a negative slope indicates the peak is behind the test point. Since the slope has the reverse polarity when the correlation value has a reverse polarity the information residing in the slope must be conditioned on the polarity of the amplitude. The amplitude conditioning of the slope is embedded in a detector S-curve formed as the product of the amplitude and the slope.

The state machine is designed to move the hypothesis test point in the direction that sets $c(\tau)\dot{c}(\tau)$ to zero. Note this happens at two locations. If $\dot{c}(\tau)$ is zero, we

Fig. 5.2 Correlation function showing slope at various test points and detector S-curve formed as product $c(\tau)\dot{c}(\tau)$

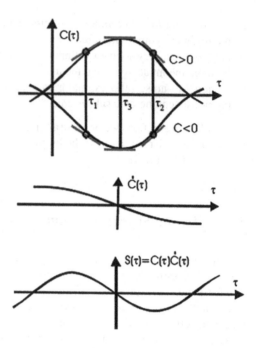

are at the peak of the correlation function and the system is a maximum likelihood estimator and if $c(\tau)$ is zero, we are at the zero of the correlation function and the system is denoted the minimum likelihood estimator.

The state machine operates like a servo system called a phase locked loop whose block diagram is shown explicitly in Fig. 5.3. The figure shows the standard components of the phase locked loop (PLL) timing loop. These include the matched filter bank, a derivative matched filter bank as well as the loop filter, which averages through the received noise and modulation noise, and the phase accumulator that selects the candidate hypothesis filter from the filter bank. Since the loop contains two integrators it is a type-2 loop, able to track a ramp input, a frequency offset, with zero steady state error.

Two other components in the phase lock loop are particularly interesting. These are the signal-to-noise ratio scale factor $2Eb/N_0$ that serves to tell the loop the quality of the signal it is processing. Remember that if the signal sample has a low SNR, the input to the loop filter should be scaled in proportion to its quality and this of course reflects the philosophy of all matched filter processes.

The second component is the $tanh$ that conditions the amplitude from the matched filter as it interacts with the derivative. At high SNR the $tanh$ defaults to the sign of the input signal as a conditional correction to the derivative information. At low SNR the $tanh$ defaults to a unity gain applied to the matched filter output to avoid possible errors in the sign decision of the amplitude as the conditional correction of the derivative. The SNR gain term in the loop filter throttles the loop bandwidth in response to the SNR. At low SNR the loop filter reduces its bandwidth

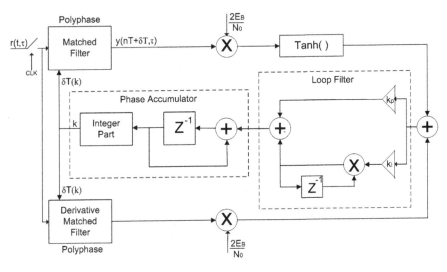

Fig. 5.3 DSP based polyphase filter bank maximum likelihood timing recovery loop

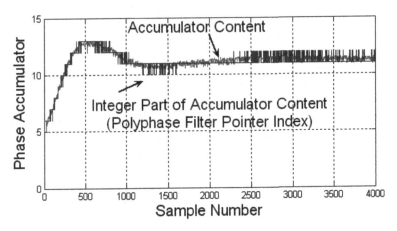

Fig. 5.4 Transient response of phase accumulator during timing acquisition

so that it has to average over a longer interval to obtain stable control signals. Conversely, at high SNR the loop filter increases its bandwidth since it can obtain stable control signals by averaging over shorter time intervals.

The problem with these two components is that most receivers do not have real time background SNR estimators operating to feed the quality assessment of data samples to the loop filter. Thus most receivers replace the *tanh* with its small SNR gain and operate as suboptimal systems.

We can obtain better results by using an M-path polyphase filter for implementing the matched filter bank. Here we only show, in Fig. 5.4, the transient response observed at the output of the phase accumulator as it moves from an initial filter path to the correct filter path in the polyphase filter bank. The red curve is the accumulator content while the blue overlay curve is the integer part of the accumulator that defines the index pointer to the selected path of the M-path filter. The interested reader can find more details on this argument in the references at the end of this chapter.

5.2.2 Phase Synchronization

In many modulators the signal formed by the shaping filter is amplitude and phase modulated in accordance with the input bit mapping process. The amplitude and phase terms are represented in Cartesian coordinates and described as a complex base band signal. The quadrature components of the signal are up converted or amplitude modulated on the quadrature components, the cosine and sine, of a radio frequency carrier.

At the demodulator the process is reversed and the radio frequency carrier is down converted by a pair of cosine and sine quadrature sinusoids. The frequency and phase of the up converter and the down converter oscillators do not match by virtue of manufacturing tolerances, age and temperature related drift, and doppler offsets due to velocity vectors between platforms.

The signal obtained at the output of the quadrature down converter is then monitored and applied to a phase detector to obtain a measure of the phase misalignment. Here we quickly describe the process by which the phase lock loop aligns the local oscillator phase with that of the phase of the received signals' underlying carrier. The key question is: what do we do with the samples of the I-Q pair, which are time aligned with the correlation peaks of their matched filters, to obtain information about their phase θ? Well, the answer is to examine a legacy solution formed by their product $I \cdot Q$. As seen in Fig. 5.5, this product is proportional to the $sin(\theta/2)$ which is the expression for an S-curve phase detector. The detector has two zero value references at $0°$ and at $180°$ which of course is responsible for the two-fold ambiguity in the synchronized phase lock.

An alternate phase detector is the $sgn(I) \cdot Q$ in which, as shown in Fig. 5.6, a more linear S-curve spans a wider range of input phase offsets. These correspond to the small signal to noise ratio and the large signal to noise ratio approximations to the $tanh$ described in the previous section. The $sgn(I) \cdot Q$ structure implements the maximum likelihood phase recovery that, under certain conditions, results in the well known Costas loop phase recovery system. The Costas loop is embedded in many legacy receivers. Here we quickly mention that it is far from an optimum phase recovery process because it performs quite poorly at low SNR.

More details on phase recovery architectures can be found in the references at the end of this chapter.

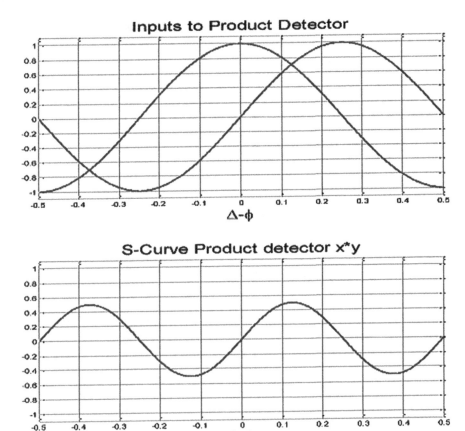

Fig. 5.5 $I \cdot Q$ BPSK phase detector and S-curve

5.2.3 Frequency Synchronization

If the frequency offset between the local oscillator used in the final down conversion and the center frequency of the input signal is sufficiently small the phase detector and the phase lock loop of the previous section can compensate it by acquiring and de-spinning the input signal. On the other hand, if the frequency offset is significantly larger than the bandwidth of the PLL loop filter the loop will not be able to successfully acquire and de-spin the signal. In this event an acquisition aid must be invoked to assist the phase lock loop.

In legacy receivers a lock detector is interrogated after a preset time-out to see if the acquisition aid should be invoked. If invoked, the local oscillator is slowly swept through the likely range of frequency offsets till the loop acquires the signal.

Inputs to Product Detector

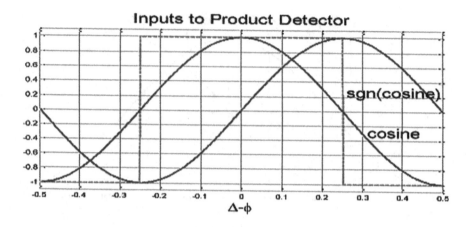

S-Curve Product Detector sign(x)*y

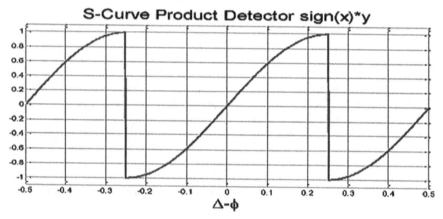

Fig. 5.6 $sgn(I) \cdot Q$ BPSK phase detector and S-curve

The acquisition is completed when the frequency offset is small enough for the frequency error signal to pass through the narrow bandwidth loop filter. The lock detector terminates the sweep upon detection of the acquisition.

In modern receivers a maximum likelihood frequency estimator performs the task of frequency acquisition. To reduce phase jitter due to the frequency lock loop noise injection, this loop is disabled when the system acquires final phase lock.

The frequency lock loop is based on a maximum likelihood frequency estimator. When we take the derivative of the output of matched filter with respect to the unknown frequency offset we obtain the frequency derivative matched filter. The frequency derivative filter, often called the band edge filter, is a sensitive detector of frequency offsets of the input signals' spectral mass; in fact when there is no offset the two symmetric band edge filters collect the same amount of signal energy. Conversely when a frequency offset is present, the two band edge filters collect different amounts of energy from the input spectrum and their energy difference contains a DC term proportional to the frequency offset. The energy difference of the band edge filters is formed as the difference of the conjugate products of the time series from each band edge filter. This difference, proportional to the frequency offset, is the input signal to the frequency lock loop-filter. A block diagram of the frequency lock loop along with its associated lock detector and its phase profile can be found in the references at the end of this chapter (see [12]).

5.3   A New Blind Synchronization Algorithm for Software Radio Receivers

After the brief introduction on time, phase and frequency synchronization in the actual digital receivers, in this section we present a novel algorithm for an ideal software radio receiver with only one analog to digital converter, placed immediately after the receiving antenna, followed by a digital data section that accomplishes the tasks of synchronizing the radio and, at same time, recovering the transmitted symbols. The algorithm presented here has a theoretical interest and it has been developed for QAM signals, further research on the same topic could work in the direction of extending it to different classes of modulated signals.

Consider a QAM signal

$$z(t) = i(t)cos(2\pi f_0 t + \theta_0) + q(t)sin(2\pi f_0 t + \theta_0) + w_g(t) \qquad (5.1)$$

where $i(t)$ and $q(t)$ are independent PAM signals with symbol rate $\frac{1}{T}$ and $w_g(t)$ is standard additive band-pass noise with power spectral density $\frac{n_g}{2}$ in the band $|f| \in (f_0 - \frac{W}{2}, f_0 + \frac{W}{2})$ with $W = \frac{2}{T}$.

The simplification in the receiver front-end and the desired flexibility pose new challenges to the design of an efficient receiver, in which most of the tasks should be assigned entirely to a numerical algorithm operating only on the discrete-time sequence $z[n]$.

More specifically:

- Since sampling is designed independently from symbol time and, the ADC is totally asynchronous from the information sequence, therefore, the samples in each symbol interval are not constant in number and they can be located in asynchronous positions (see Fig. 5.7 for a simplified case).
- The receiver may have no exact knowledge of the carrier frequency.
- The receiver has no knowledge of the phase.
- The noise is slightly correlated.
- Computational complexity must be reasonable.

Figure 5.8 shows our proposed receiver where, after the band-pass filter and the ADC, we have included a counter that tracks the number of samples in each symbol interval. We assume that the symbol time T and the sampling time T_s are known and the second one is not necessarily a multiple of the first one.

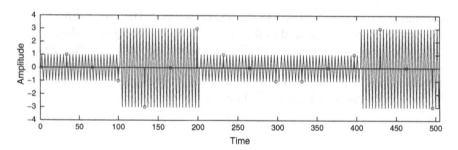

Fig. 5.7 A segment of sparse sampled QAM signal

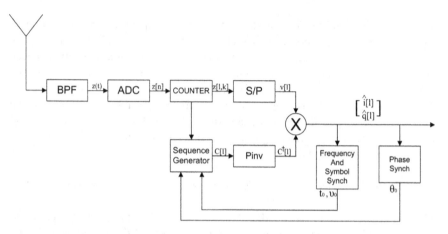

Fig. 5.8 Block diagram of the proposed receiver

The samples occurring in the lth symbol interval are organized in the vector

$$v[l] = \begin{pmatrix} z[l, k]_{k=1}^{N-1} \\ z[l, N] \end{pmatrix} = \begin{pmatrix} z\left[k + l(N-1) + \sum_{i=0}^{l} g_{i-l}\right]_{k=1}^{N-1} \\ g_l z\left[N + l(N-1) + \sum_{i=0}^{l} g_{i-l}\right] \end{pmatrix} \quad (5.2)$$

where at each symbol interval a counter generates a value $g_l = 1$ if the number of samples is N, otherwise $g_l = 0$ if the number of samples is $N - 1$, in which case the last element is null.

The relation with the transmitted symbols and the noise is

$$v[l] = C[l] \begin{bmatrix} i[l] \\ q[l] \end{bmatrix} + w_g[l] + w_a[l] \quad (5.3)$$

where $w_g[l]$ and $w_a[l]$ are windows of noise samples and the $N * 2$ matrix $C[l]$ contains the known signal samples

$$C[l] = \begin{pmatrix} a[l,k]_{k=1}^{N-1} & b[l,k]_{k=1}^{N-1} \\ g_l a[l,N] & g_l b[l,N] \end{pmatrix} \tag{5.4}$$

$$a[l,k] = \cos\left[2\pi\nu_0\left(k + l(N-1) + \sum_{i=0}^{l} g_{i-l} + \frac{t_0}{T_s}\right) + \theta_0\right] \tag{5.5}$$

$$b[l,k] = \sin\left[2\pi\nu_0\left(k + l(N-1) + \sum_{i=0}^{l} g_{i-l} + \frac{t_0}{T_s}\right) + \theta_0\right]. \tag{5.6}$$

The noise vectors are N-dimensional with the last element null if $g_l = 0$ and total autocorrelation matrix for $w = w_g + w_a$,

$$R_w = Toeplitz\left(r_a[0] + r_g[0] + r_g[1] + \cdots + r_g[N-2] + 0\right) \tag{5.7}$$

for the windows with $N - 1$ samples and

$$R_w = Toeplitz\left(r_a[0] + r_g[0] + r_g[1] + \cdots + r_g[N-1]\right) \tag{5.8}$$

for the windows with N samples.

The optimum receiver for perfectly known parameters t_0, ν_0, θ_0, $r_g[l]$ and $r_a[0]$, because of the correlated noise, would require the first stage with a whitening matrix, followed by a minimum distance classifier. Equivalently a zero-forcing matrix followed by a Mahalanobis distance classifier determined by the noise covariance matrix eigenvalue spread, would also be optimal.

We consider here only a suboptimal minimum mean square error receiver based on the assumption that the noise contribution is gaussian and white

$$\begin{pmatrix} \hat{i}[l] \\ \hat{q}[l] \end{pmatrix} = C^\dagger[l]v[l] \tag{5.9}$$

where $C^\dagger$ is the pseudo-inverse matrix of $C[l]$.

The independence assumption on the noise components becomes closer to reality as the digital bandwidth Δ approaches 0.5 (larger values of n). On the other hand it is unlikely that the ratio between $\eta_a[0]$ and $\eta_b[0]$ is known, or that it can be estimated with precision. Therefore a good design, that keeps into account the slight correlation among the noise components, may not be too robust anyways.

Note that the receiver is time-varying and the pseudo-inverse must be calculated at every symbol interval. This is a consequence of the total asynchronism of symbol and sampling times.

5.3.1 Symbol, Frequency and Phase Synchronization

In perfect synchronization conditions, the receiver output should see the initial constellation on $(i[l], q[l])^T$ with each signal point scattered in a gaussian cloud. Each cloud is practically circular because of very slight residual noise correlation after filtering. Conversely, if the receiver has no knowledge of t_0 and θ_0, and imprecise knowledge of ν_0, the output will look confusing.

We propose here a blind algorithm that locks in to the correct parameters by verifying in real-time the regularity of the constellation.

Assuming a typical rectangular constellation for $(i[l], q[l])$, we have successfully used maximization of an on-line estimate of the joint entropy of $(i[l]^\dagger, q[l]^\dagger)$ to recover frequency ν_0 and time synchronization t_0 jointly.

It is well known that the uniform distribution maximizes the joint entropy of two random variables. In the case of interest the symmetry of the rectangular constellation, even in the presence of noise, allows our algorithm based on entropy maximization, to pick up the right parameters sorting out all the confused joint distributions that emerge in the absence of synchronization. Once time and frequency synchronization has been achieved, the constellation appears at the output rotated by an angle equal to the phase θ_0. Minimizing the mutual information of $i^\dagger[l]$ and $q^\dagger[l]$ we can recover θ_0 with good precision, except for the obvious residual rotations in multiples of $\frac{\pi}{2}$.

The proposed algorithm follows the footsteps of the many information-theory based algorithms proposed in the literature for blind signal separation. In our case the scenario is much simpler because the signal space is only two-dimensional and it is not too complex to use histograms and exhaustive searches.

The entropy and mutual information estimates are based on a two-dimensional $N_b * N_b$ histogram computed on a time frame of L symbols. The elements p_{nm} of the histogram are defined as the normalized number of occurrences of $(i[l]^\dagger, q[l]^\dagger)$ falling into the generic bin $[(n-l)\Delta, n\Delta] * [(m-l)\Delta, m\Delta]$. In the computation to each histogram's element we add a spare occurrence to each bin to avoid the numerical indeterminacies in the computation of the logarithm. The marginal distributions are

$$P_{in} = \sum_m P_{nm} \qquad P_{qn} = \sum_n P_{nm}. \tag{5.10}$$

Therefore ν_0 and t_0 are obtained by an exhaustive search on discrete values of

$$H_{iq} = -\sum_{n,m} P_{nm} log(P_{nm}) \tag{5.11}$$

and similarly the phase θ_0 from $I_{iq} = H_i + H_q - H_{iq}$ with

$$H_i = -\sum_n P_{in} log(P_{in}) \qquad H_q = -\sum_m P_{qm} log(P_{qm}) \tag{5.12}$$

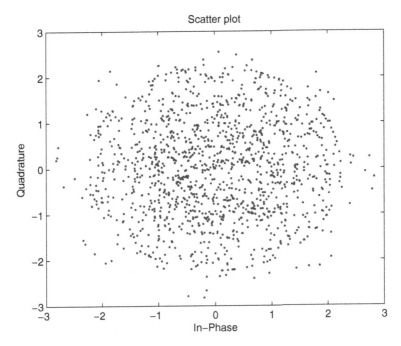

Fig. 5.9 Scatter plot of $(i^\dagger[l], q^\dagger[l])$ without synchronization

5.3.2   *Simulation Results*

We report here the result of some simulations of a QAM receiver with $f_0 = 21,500\,\text{Hz}$, $T = 8 * 10^{-4}\,s$, $W = 3\,\text{MHz}$ and sampling rate $f_s = 6,619\,\text{Hz}$ with $n = 7$. Here the constellation is a 16-QAM and the band-pass noise is Gaussian with $\eta_0 = 0.5$ and zero mean, also the devices' noise is Gaussian with $\eta_a = 0.005$ and zero mean. In the numerical frequency domain after band-pass sampling the bandwidth of sampled signal is $\Delta = 0.4530$.

Figure 5.9 depicts a scatter plot of $(i^\dagger[l], q^\dagger[l])$ without any synchronization while Fig. 5.10 represents the result of joint entropy maximization.

Figure 5.11 is a plot of the joint entropy as a function of time delay t and frequency range ν with a clear global maximum at t_0 and ν_0.

In these simulations the number of bins for the histogram is $N_b = 70$. The order of the histogram can be chosen as a compromise between precision and computational complexity. The number of symbols for estimating the histogram for every value of t and ν is $L = 150$. The parameters t and ν have been respectively spaced with a step equal to $\frac{T}{100}$ and $\frac{1}{f_s}$.

Figure 5.12 shows the scatter plot after phase synchronization and Fig. 5.13 shows a plot of the mutual information as a function of the rotation angle, sampled

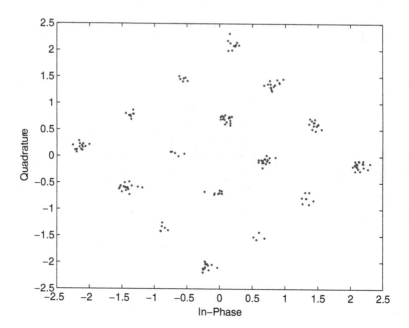

Fig. 5.10 Scatter plot of $(i^\dagger[l], q\dagger[l])$ after time and frequency synchronization

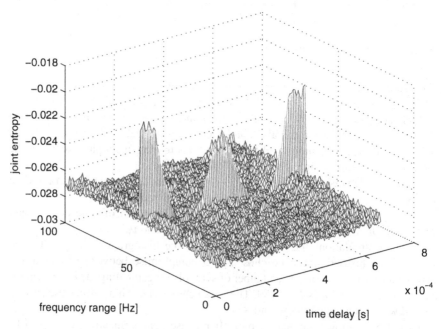

Fig. 5.11 Joint entropy of $(i^\dagger[l], q\dagger[l])$ as a function of time delay t and frequency νf_s (correct values are $t_0 = T = 8 * 10^{-4}$ s and $\nu_0 f_s = 50$ Hz)

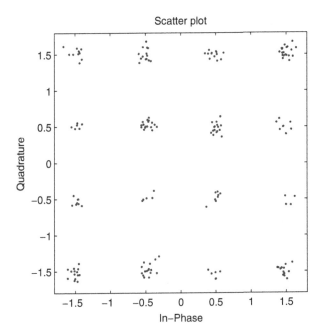

Fig. 5.12 Scatter plot of $(i^\dagger[l], q^\dagger[l])$ after phase synchronization

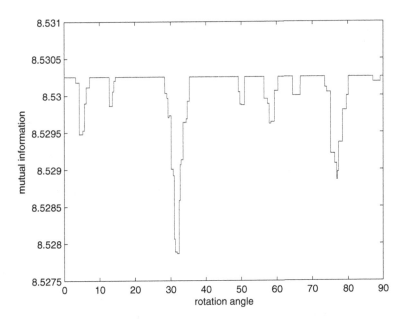

Fig. 5.13 Mutual information of $(i^\dagger[l], q^\dagger[l])$ as a function of the rotation angle (correct value is $\theta_0 = 32°$)

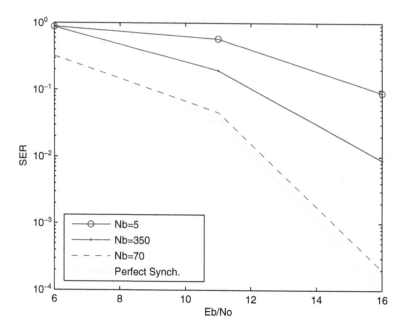

Fig. 5.14 Symbol error rate of a 16-QAM constellation for several values of N_b

with a step of $1°$. Clearly the constellation is well recovered with the obvious uncertainty of multiples of $90°$ that must be solved by other methods (as in the most blind methods).

To evaluate also how critical the choice of N_b and L is, Fig. 5.14 shows the typical symbol error rate of a 16-QAM constellation for several values of N_b and $L = 150$. It's clear from the figure that $N_b = 5$ is not sufficient number of bins while $N_b = 350$ is too high for only 150 symbols. $N_b = 70$ seems to be the best choice.

The last curve in Fig. 5.14 represents the symbol error rate (SER) of our receiver assuming perfect synchronization. Several simulations were performed with different number of symbols and constellations in order to support these results and they all confirm this typical behavior.

5.4 Recap

At the beginning of this chapter we presented a very brief overview of the synchronization process performed in today's digital radio communication systems. The particular processes we examined were timing recovery, phase recovery, and frequency recovery. We tried to avoid mathematical derivations but rather to emphasize concepts. More information on this topic can be found in the references proposed at the end of the chapter.

In this same chapter we also presented a blind algorithm for synchronizing a fully-digital QAM receiver using band-pass sampling as the first stage. That avoids the analog heterodyne, or intermediate frequency conversion usually performed in the front-end receiver. The interest of the proposed algorithm is more theoretical than practical, but it can give good suggestions for advanced research in this same area. Because the sampling rate of the only ADC present in the receiver has been chosen according to the theory of band-pass sampling, the receiver design is carried out with reference to the noise aliasing caused by the hardware devices.

The proposed algorithm is based on maximization of joint entropy and minimization of mutual information of two recovered digital sequences. Time, frequency and phase synchronization is simultaneously obtained.

Success of the method suggests further analysis of computational complexity of inversion and synchronization processes and more simulations on experimental channel models.

References

1. E. Buracchini, CSELT, "The Software Radio Concept," *IEEE Communications Magazine*, vol. 38, no. 9, 2000.
2. Rodney G. Vaughan, Neil L. Scott, and D. Rod White, "The Theory of Bandpass Sampling," *IEEE Transaction on Signal Processing*, vol. 39, no. 9, 1991.
3. Jean-Francois Cardoso, C.N.R.S and E.N.S.T, "Blind Signal Separation: Statistical Principles," *Proceedings of the IEEE*, vol. 9, no. 10, 1998.
4. D. A. Linden, "A Discussion of Sampling Theorem," *Proceedings of the IRE*, vol. 47, no. 7, 1959.
5. G. Guirong, Z. Zhaowen, W. Feixue, "Mixer-Free All Digital Quadrature Demodulation," *Proceedings of the ICSP*, Beijing, China, 12-16 October, 1998.
6. David Tse, Pramod Viswanath, *Fundamentals of Wireless Communications*, New York, Cambridge University Press, 2005.
7. Thomas M. Cover, Joy A. Thomas, *Elements of Information Theory*, New York, Wiley, 1991.
8. Adrian Karl Ong, *Bandpass Analog-to-Digital Conversion for Wireless Applications*, Ph.D. thesis, Stanford University, 1998.
9. H. L. Van Trees, Detection, *Estimation and Modulation Theory*, New York, John Wiley and Sons, 2001.
10. L. P. Sabel, D. Tran, "An Analysis of IF Filter in Bandpass Sampling Digital Demodulators," *Radio Receivers and Associated Systems*, vol. 2, no. 415, 1995.
11. Yi-Ran Sun, *Generalized Bandpass Sampling Receivers for Software Defined Radio*, Ph.D. thesis, Royal Institute of Technology, Stockholm, 2006.
12. fred harris, "Lets Assume the System is Synchronized," Monograph, Ramjee Prasads 50th-Ph.D. Celebration, Aalborg, Denmark, 11 April, 2008.
13. M. Rice, C. Dick, and f. harris, "Maximum Likelihood Carrier Phase Synchronization in FPGA-Based Software Defined Radios," *IEEE International Conference on Acoustics, Speech, and Signal Processing,* vol. 2, pp. 889892, May 2001.
14. Chris Dick, fred harris, Michael Rice, "Synchronization in Software Radios - Carrier and Timing Recovery Using FPGAs," *IEEE Symposium on Field-Programmable Custom Computing Machines,* 2000.

15. f. j. harris and M. Rice, "Multirate Digital Filters for Symbol Timing Synchronization in Software Defined Radios," *IEEE Journal on Selected Areas of Communications,* vol. 19, no. 12, December 2001.
16. Chris Dick, Benjamin Egg, fred harris, "Architecture and Simulation of Timing Synchronization Circuits for the FPGA Implementation of Narrowband Waveforms," *Proceeding of the SDR'06 Technical Conference and Product Exposition,* 2006.
17. fred harris, "Band Edge Filtering and Processing for Timing and Timing Recovery," *ComCon-7,* Athens, Greece, 28-June, 2-July 1999.

Index

E. Venosa et al., *Software Radio: Sampling Rate Selection, Design and Synchronization,* 129
Analog Circuits and Signal Processing, DOI 10.1007/978-1-4614-0113-1,
© Springer Science+Business Media, LLC 2012